# Geochemistry and Health

T0258831

*Edited by*

Iain Thornton with assistance from Hazel Doyle and Ann Moir

*Centre for Environmental Technology*
Imperial College of Science, Technology and Medicine
London

## CRC Press
Taylor & Francis Group
Boca Raton London New York

CRC Press is an imprint of the
Taylor & Francis Group, an **informa** business

# Contents

# 1 The Geochemical Atlas as a Means to Establishing the Balance of Nutritional Elements in Finnish Soils

H. Tanskanen, N. Gustavsson, T. Koljonen and P. Noras
*Geological Survey of Finland*
*Kivimiehentie 1, SF-02150 Espoo, Finland*

## Summary

*In 1984-85 the Geological Survey of Finland carried out countrywide geochemical mapping of till at 1:2,000,000 to establish the regional contents and behaviour of major and trace elements and to provide material for the first Geochemical Atlas of Finland. Till was found to be a very satisfactory study material because the sedimentary cover in Finland is usually thin (1-10 m), the unconsolidated sediments are young (<10,000 y) and they have been transported only a short distance from the place of origin. We believe that the methods used can be applied in most parts of the world where sediments are of local origin, and even in tropical areas, where laterite can serve as the sampling material. It is proposed that low density geochemical mapping should be done wherever possible, to establish the balance of the key nutritional elements on the earth's surface.*

## Sampling

The primary data for the Atlas is provided by 1,057 composite till samples collected from all over Finland. The mean sampling site density was one sample per 300 km$^2$. Each sample is composite and comprises five subsamples, collected within an area of 0.5 by 2.5 km$^2$. The samples (ca. 0.5 kg) were dried and sieved through a 0.06 mm screen.

## Analytical Methods

The fine fraction of till was analyzed by multi-element techniques, *i.e.* plasma emission spectrometry (ICP-AES) and epithermal neutron activation analysis (ENAA). Other determinations were carried out by electrothermal atomic absorption

1

Table 1. *Elements analyzed and the lowest determination limit (ppm) of the chemical analyses*

| Elements | ICP-AES (a) | (b) | ENAA | Other | Elements | ICP-AES (a) | (b) | ENAA | Other |
|---|---|---|---|---|---|---|---|---|---|
| Ag |  | 2* |  |  | Na | x | x |  |  |
| Al | x | x |  |  | Ni | 10 | 5 |  |  |
| As |  |  | x |  | P | x | x |  |  |
| Au |  |  |  | 0.001 | Pb |  | 10 |  |  |
| Ba | x | x |  |  | Pd |  |  | x | 0.001 |
| Br |  |  | 1.5 |  | Rb |  |  | x |  |
| Ca | x | x |  |  | S |  |  |  | 20 |
| Cd |  | 2 |  |  | Sb |  |  |  | 0.2 |
| Co | 5 | 2 |  |  | Sc | 3 | 1 |  |  |
| Cr | 10 | 5 |  |  | Si | x |  |  |  |
| Cs |  |  | 1.5 |  | Sm |  |  |  | 1.5 |
| Cu | 5 | 2 |  |  | Sr | x | x |  |  |
| Fe | x | x |  |  | Ta |  |  |  | 1 |
| K | x | x |  |  | Th |  | 20 |  |  |
| La | 10 | 5 |  |  | Ti | x | x |  |  |
| Li | 5 | 2 |  |  | U |  |  | x |  |
| Lu |  |  | 0.1 |  | V | x | x |  |  |
| Mg | x | x |  |  | Y | 5 | 2 |  |  |
| Mn | x | x |  |  | Zn | 5 | 2 |  |  |
| Mo |  | 1 |  |  | Zr | x |  |  |  |

(a) total dissolution; (b) partial leach.
 x determination limit is much below the concentrations normally present in till.
2* 2 ppm determination limit. The between-batch precision is poor for Ag, Cd, Mo and Pb.

spectrometry (Au and Pd), and sulphur was determined with a Leco Analyser. So far 40 elements have been determined (Table 1).

Both total decomposition (a) and partial leach (b) were used in the preparation of the sample solutions (Table 1). Total decomposition involves digesting with cold concentrated hydrofluoric acid, heating with aqua regia and neutralising with boric acid. The HF procedure gives total concentrations of elements for most samples and the ENAA technique for all samples. The partial leach is based on digestion with hot concentrated aqua regia, which only dissolves most of the non-silicates and many of the looser silicate structures.

**Table 2.** *The rock groups presented in scattergrams in Figures 1 and 3. The grouping is based on the geological map by Simonen (1980).*

---

Postsvecokarelian (1540-1679 My):
10  Rapakivi granite

Svecokarelian (1800-1900 My):
13  Granite
14  Granite veins in migmatitic gneis
15  Granodiorite and quartz diorite
16  Gabbro, anorthosite and peridotite
18  Metabasalt, greenstones and amphibolite
19  Phyllite, mica schist and mica gneiss
20  Quartz-feldspar schist and gneiss
21  Quartzite

Presvecokarelian igneos rocks, younger than basement complex
(ca. 2440 My):
22  Peridotite, gabbro and anorthosite

Presvecokarelian basement complex (2500-3100 My):
24  Gneiss of the granulite complex
27  Granitic veins in basement gneiss
31  Mica  schist and mica gneiss
32  Quartz-feldspar schist and gneiss

---

### Data Processing

The processing of the Atlas data is done in the normal way for the stages of quality control, data storage, statistical summary and display on maps (Gustavsson *et al.*, 1987). The basic type of presentation is a dot map with dot size as the indicator of contents (Figures 2 and 4). The dot size is exponentially related to the distribution of the variable, which gives emphasis to the upper tail of the distribution (Björklund and Gustavsson, 1987). Only relevant variations at levels exceeding the analytical detection limit are visibly distinguished.

The raster technique is employed to achieve distinct black dots with erasive white frames showing overlapping dots. Stored raster images, such as borders and other maps, can be drawn by masking or, in the case of the background for the dot map, by overlaying.

The legend of the dot map shows the cumulative distribution function of the variable together with the dot size function curve. A sample of dots is drawn along the concentration axis to aid eyeball estimation of single dots.

3

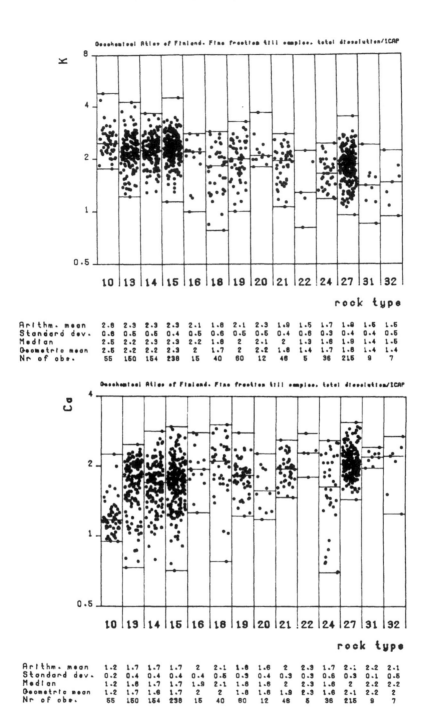

| | | | | | | | | | | | | | | |
|---|---|---|---|---|---|---|---|---|---|---|---|---|---|---|
| Arithm. mean | 2.6 | 2.3 | 2.3 | 2.3 | 2.1 | 1.8 | 2.1 | 2.3 | 1.9 | 1.5 | 1.7 | 1.8 | 1.5 | 1.5 |
| Standard dev. | 0.8 | 0.5 | 0.5 | 0.4 | 0.5 | 0.6 | 0.5 | 0.4 | 0.6 | 0.3 | 0.4 | 0.4 | 0.5 |
| Median | 2.5 | 2.2 | 2.3 | 2.3 | 2.2 | 1.8 | 2 | 2.1 | 2 | 1.3 | 1.6 | 1.9 | 1.4 | 1.5 |
| Geometric mean | 2.5 | 2.2 | 2.2 | 2.3 | 2 | 1.7 | 2 | 2.2 | 1.8 | 1.4 | 1.7 | 1.8 | 1.4 | 1.4 |
| Nr of obs. | 55 | 150 | 154 | 238 | 15 | 40 | 60 | 12 | 46 | 5 | 36 | 215 | 9 | 7 |

| | | | | | | | | | | | | | | |
|---|---|---|---|---|---|---|---|---|---|---|---|---|---|---|
| Arithm. mean | 1.2 | 1.7 | 1.7 | 1.7 | 2 | 2.1 | 1.8 | 1.6 | 2 | 2.3 | 1.7 | 2.1 | 2.2 | 2.1 |
| Standard dev. | 0.2 | 0.4 | 0.4 | 0.4 | 0.4 | 0.5 | 0.3 | 0.4 | 0.3 | 0.3 | 0.6 | 0.3 | 0.1 | 0.5 |
| Median | 1.2 | 1.6 | 1.7 | 1.7 | 1.9 | 2.1 | 1.8 | 1.6 | 2 | 2.3 | 1.6 | 2 | 2.2 | 2.2 |
| Geometric mean | 1.2 | 1.7 | 1.6 | 1.7 | 2 | 2 | 1.8 | 1.6 | 1.9 | 2.3 | 1.6 | 2.1 | 2.2 | 2 |
| Nr of obs. | 55 | 150 | 154 | 238 | 15 | 40 | 60 | 12 | 46 | 5 | 36 | 215 | 9 | 7 |

**Figure 1.** *Scattergrams showing the contents of K and Ca in till (-62 μm fraction) in areas differing in main rock type. The rock types are based on the geological map of Simonen (1980) and are explained in Table 2 (nos.10, 13, ... 32). Only the most common rock types are shown.*

**Table 3.** *The average contents (%) and ratios of selected elements: (\*), in bedrock in a granitoid area (granite 34.0 vol.%, quartz diorite and granodiorite 62.1 vol.%, gabbro 3.6 vol.% and peridotite 0.3 vol.%), in till in a granodiorite-quartz diorite area ("rock type" code 15, Table 2, Figures 1 and 3) and in the till of all Atlas samples.*

|      | (*) | Finnish till (all Atlas data) | | | |
|------|-----|-------|------|-----------|--------|
|      |     | No.15 | Mean | Std. dev. | Median |
| K    | 3.1 | 2.3   | 2.1  | 0.5       | 2.1    |
| Ca   | 2.2 | 1.7   | 1.8  | 0.4       | 1.8    |
| Mg   | 0.8 | 0.9   | 1.0  | 0.4       | 1.0    |
| Ca/K | 0.7 | 0.8   | 0.9  | 0.4       | 0.9    |
| Ca/Mg| 2.7 | 2.1   | 1.9  | 0.7       | 1.8    |
| K/Mg | 3.9 | 2.9   | 2.5  | 1.7       | 2.2    |

(*) Koljonen and Carlson 1975, Table 1

The rock type is derived from the geological map of Finland by Simonen (1980), using a 5 km x 5 km squared grid and estimating the dominating rock type for each grid cell. Each sample is labelled with the bedrock type of the cell containing the sampling site.

**Results**

The contents of K and Ca and the ratios of Ca/K and Ca/Mg in till are shown as scattergrams in Figures 1 and 3 and the distributions of K, Ca, Ca/K and Ca/Mg as maps in Figures 2 and 4. The results are presented in the form these are available at the Geological Survey of Finland. The maps are also produced as surface maps with greytone or colour scale as indicators (Gustavsson *et al.*, 1987), and will be published in Atlas form in the near future.

A good correspondence can be seen between the elemental concentrations in till and the bedrock type, *e.g.* the Ca/K ratio is high in northern Finland where basaltic lavas prevail and low in southern Finland where granitoid rocks prevail (Figure 4). As expected, the chemical differences in the bedrock are masked due to mixing in till (Table 3). Till as a sampling media has the advantage that it more accurately represents the mean chemical composition of bedrock than rock samples. This results from difficulties in having representative rock samples where the variations in bedrock are great. Variations are expected especially in southern Finland where granitic veins occur in migmatitic gneisses and schists. The results presented in the figures and tables are in accordance with the known behaviour of K, Ca and Mg in

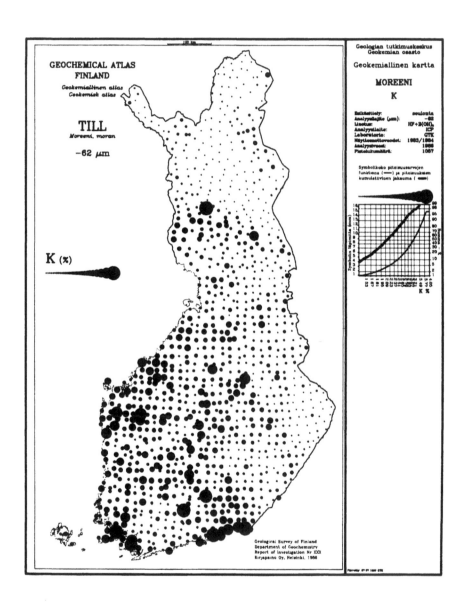

**Figure 2.** *The geochemical maps of K and Ca in till (-62 μm fraction). Original scale 1:2,000,000.*

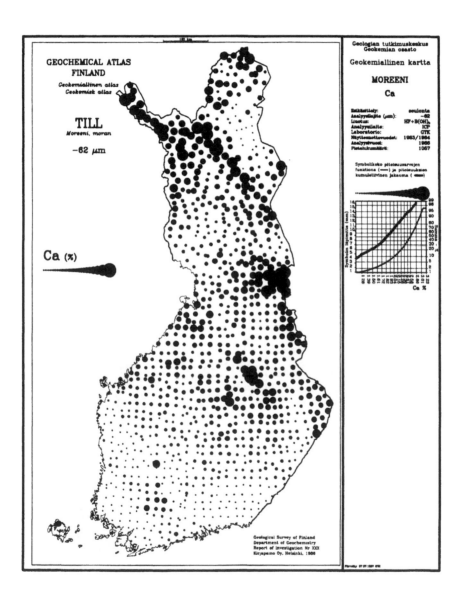

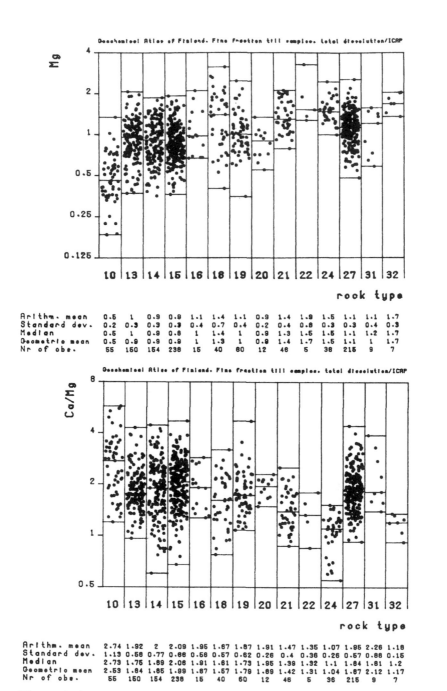

**Figure 3.** *Scattergrams showing the content of Mg and the Ca/Mg ratio in till (-62 μm fraction) in areas differing in main rock type. The rock types are based on the geological map of Simonen (1980) and are explained in Table 2 (nos.10, 13, ... 32). Only the most common rock types are shown.*

8

exogenic processes, but the Atlas-type of data provides numerical values, more exact than any available earlier. With these more appropriate data, separate areas can be compared and the migration and especially the relative migration of chemical elements can be followed.

Table 3 shows the K content to be highest (K 3.1%) in the granitoid rocks. Although it decreases (K 2.3%) in the till collected from granodiorite-quartz diorite areas (these rocks being more basic than granite, form together with granite the granitoid group), it is still higher than the average (K 2.1%) in Finnish till.

The content of Ca, which in bedrock is enriched in basic rocks, is low (Ca 2.2%; Ca/K 0.7) in granitoid rocks and even decreases in till in granodiorite-quartz diorite areas (Ca 1.7%; Ca/K 0.8). This last value of Ca is nearly the same as the average of Finnish till (Ca 1.8%). This suggests that in Finnish soils Ca migrates faster than K, which is adsorbed onto clays.

In endogenic processes the behaviour of Mg resembles that of Ca. Both elements are enriched in basic rocks, but in exogenic processes Mg migrates more slowly than Ca, as is indicated by the significant decrease in Ca/Mg ratios in Table 3 from bedrock (2.7) to till (2.1 and 1.9). The slower migration is probably due to the collection of Mg in clay minerals and chlorite in soils, as proposed by Koljonen and Carlson (1975). Calcium carbonate, which could incorporate Ca in an immobile form, does not precipitate because Finnish soils are acid and podzolized.

## Conclusions

The results illustrated in the Atlas maps are in accordance with the known geochemistry of the elements studied. This suggests that the methods employed (sampling, chemical analyses and data processing) are justified and that regional, national geochemical mapping is a useful tool for studying the abundance and behaviour of elements on the surface and establishing the balance of nutritional elements in soils. The areal features are well displayed with the sampling density used. More detailed information requires a higher density, and in fact all of Finland is to be geochemically mapped with a density of 1 sample per 4 km$^2$ by the end of 1991.

## References

Björklund, A. and Gustavsson, N. (in press 1987). Visualisation of geochemical data as maps. *J. Geochem. Expl.*, 27.

Gustavsson, N., Koljonen, T., Noras, P. and Tanskanen, H. (in press 1987). Geochemical Atlas of Finland. *J. Geochem. Expl.*

Koljonen, T. and Carlson, L. (1975). Behaviour of major elements and minerals in sediments of four humic lakes in south-western Finland. *Fennia*, 137, 47.

Simonen, A. (1980). Prequaternary Rocks of Finland. Geological map, 1:1,000,000. Geological Survey of Finland, Espoo.

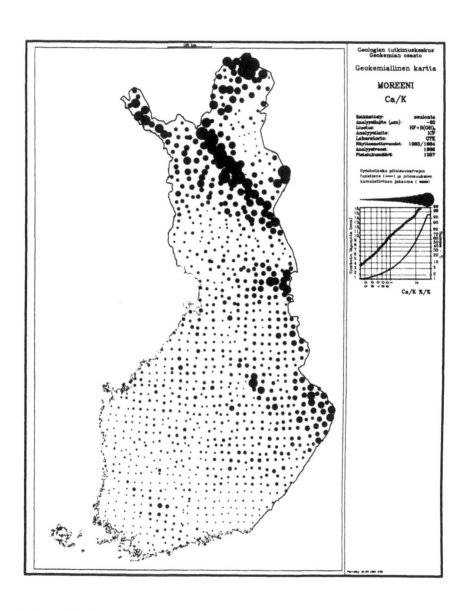

**Figure 4.** *The geochemical maps of the Ca/K and Ca/Mg ratios in till (-62 μm fraction). Original scale 1:2,000,000.*

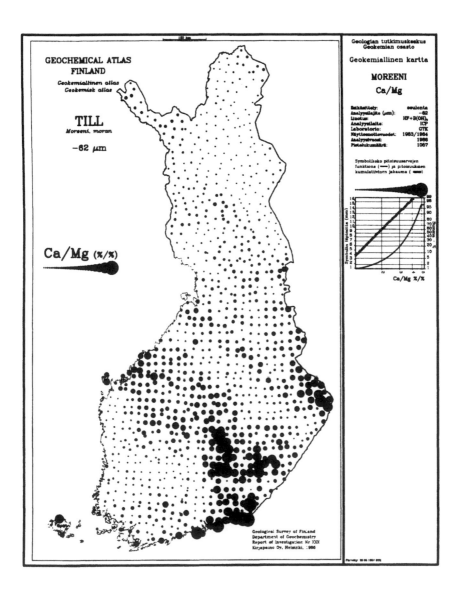

GEOCHEMICAL ATLAS
FINLAND
*Geokemiallinen atlas*
*Geokemisk atlas*

TILL
*Moreeni, moran*

−62 μm

Ca/Mg (%/%)

Geologian tutkimuskeskus
Geokemian osasto

Geokemiallinen kartta

MOREENI
Ca/Mg

| | |
|---|---|
| Esikäsittely: | seulonta |
| Analyysilajike (μm): | −62 |
| Liuotus: | HF+B(OH)₃ |
| Analyysilaite: | ICP |
| Laboratorio: | GTK |
| Näytteenottovuodet: | 1983/1984 |
| Analyysivuosi: | 1986 |
| Pistehukumäärä: | 1067 |

Symbolikoko pitoisuusarvojen
funktiona (───) ja pitoisuuksien
kumulatiivinen jakauma (═══)

Ca/Mg %/%

Geological Survey of Finland
Department of Geochemistry
Report of Investigation Nr XXX
Kirjapaino Oy, Helsinki, 1986

11

# 2 Biogeochemical mapping of Sweden for geomedical and environmental research

O.Selinus
*Geological Survey of Sweden, Geochemical Department*
*PO Box 670, S-75128 Uppsala, Sweden*

## The Geochemical Mapping Programme

The Geological Survey of Sweden started a national mapping programme in 1982. The purpose of this programme is to compile a geochemical atlas of the entire country. The maps outline the broad-scale patterns for heavy metals in the surface environment. A new method has been used in this programme. Instead of sampling inorganic stream sediments we use organic material consisting of plant roots from stream banks and aquatic mosses. These reflect in a very sensitive way the contents of heavy metals in stream water. We have also seen that roots and mosses both respond closely to chemical variations at the background levels related to different bedrock types. They also reflect in a sensitive way the effects of pollution.

The mapping programme also includes soil surveys (0.16 samples per $km^2$).

## The Sampling Material

Aerial parts of many plant species do not, in general, respond to high heavy metal concentrations in the growth medium. This is due to the fact that plants have physiological barriers between roots and above-ground parts protecting them from uptake of toxic levels of heavy metals into the vital reproductive organs. Aerial parts of plants may however respond to the availability of heavy metals in the growth medium provided the metal concentrations are below the physiological barriers. This means that sampling of aerial parts of plants are not usually suitable for geochemical prospecting or mapping. Roots and aquatic mosses are however suitable (Brundin *et al.*, 1987). Therefore, we consider this sampling medium to be the best for our mapping programme. Studies have also shown that roots and mosses can be used in other, widely different climatic conditions, *e.g.* in the tropics.

All sampling points are chosen in such a way that they each represent a drainage area (0.16 samples per $km^2$).

13

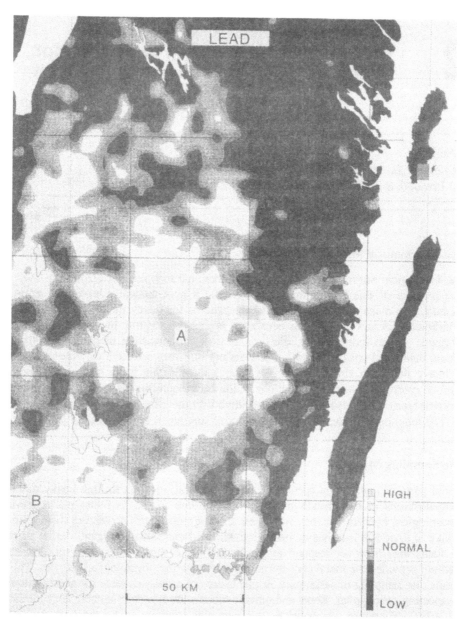

**Figure 1.** *Biogeochemical map of southern Sweden. Lead in organic stream bank material i.e. plant roots and aquatic mosses. 5442 samples. Colours indicate values in relation to the whole country.*

14

After analysis by X-ray fluorescence (XRF), atomic absorption (AA) and neutron activation (NAA) all analytical results of 32 elements are processed by computer.

The geochemical patterns have been smoothed in the colour maps shown by the Kriging method. Each colour represents a percentile interval of 10%. Before Kriging, all raw-values are normalised with respect to organic content and limonite content using stepwise linear regression (Selinus, 1983). The colour maps therefore show the metal contents relative to the average contents in Sweden. Of course the procedure is completely computerised and all data are stored in data bases making it easy for medical research centres, agencies and companies to get the data the way they want.

## Monitoring and Bedrock Surveys

Monitoring surveys are carried out at regular time intervals. Thirty sampling stations represent background conditions in different geological, climatological and physiographical environments in Sweden. At these stations samples are taken from vegetation, the soil-horizons, groundwater and drainage sediments. The results from these stations can be used for the interpretation of the broad-scale geochemical patterns. They also offer an opportunity to detect long term and short term changes in the geochemical environment and the amount of airborne pollution (by means of forest moss).

Bedrock surveys are also carried out. Samples of all bedrock types from the mapped areas are analysed and by means of multivariate statistical methods we can interpret the maps and distinguish between natural background and antropogenic sources.

## Examples of Maps

The general trend of lead is the enhanced lead contents to the south that have a decreasing tendency towards the north (Figure 1). This lead pattern is probably derived from airborne lead contamination from in part the industrialised areas on the continent and Great Britain. It also shows the effects of acid rain on the mobilisation of heavy metals. In the central part there is a distinct maximum (A) which coincides with an area where several glass manufacturing industries are located. The high lead values of area B are caused by the bedrock which consists of ore bearing precambrian volcanics.

The bedrock of Sweden consists mainly of precambrian granites, gneisses and volcanic rocks. These are almost deficient in selenium which can be seen in the northern part of the map (A, Figure 2). The most southern part of Sweden however consists of mesozoic sediments. These have generally higher contents of selenium (B). This is fortunate since one of Sweden's most intensive agricultural areas is located here.

The very high chromium contents in this area (A, Figure 3) are caused by the dolerites and metabasites which are rich in this element. This is a good example of the heavy impact natural sources can have on the environment, and the importance of geochemical maps in medicine.

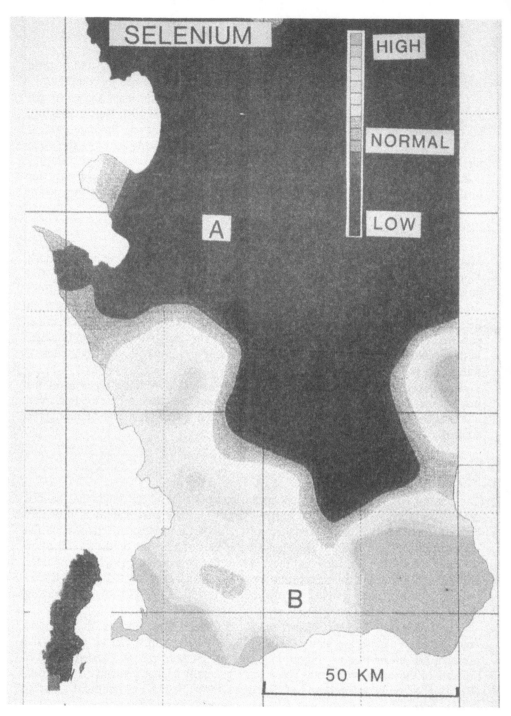

**Figure 2.** *Biogeochemical map of southern Sweden. Selenium in organic stream bank material i.e. roots and aquatic mosses. 442 samples. Colours indicate values in relation to the whole country.*

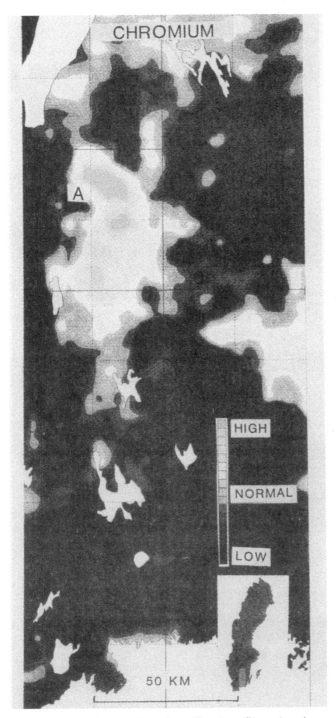

**Figure 3.** *Biogeochemical map of southern Sweden. Chromium in organic stream bank material i.e. plant roots and aquatic mosses. 3952 samples. Colours indicate values in relation to the whole country.*

17

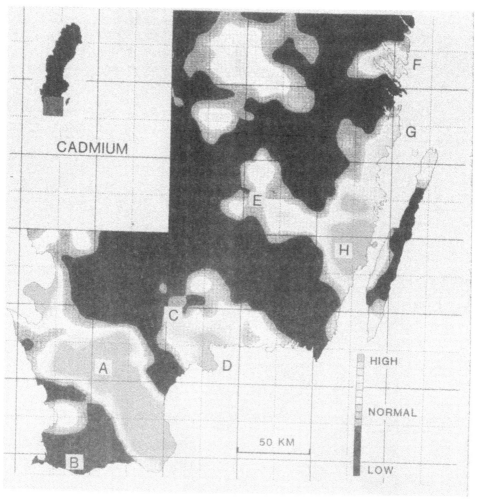

**Figure 4**. *Biogeochemical map of southern Sweden. Cadmium in organic stream bank material i.e. plant roots and aquatic mosses. 1248 samples. Colours indicate values in relation to the whole country.*

The high cadmium contents in area A (Figure 4) are most certainly caused by a combination of pollution by air emission from the continent and the use of fertilisers and sewage sludge in agriculture. Area B is however low in cadmium although this is also an intensive agricultural area. The reason for this is the bedrock consisting of mainly mesozoic limestones causing this area to have a great buffering capacity against acid rain. Therefore the mobilisation of cadmium is greater in area A than in area B. The bedrock in area C consists mainly of proterozoic volcanites with some mineralisations. Therefore the cadmium contents are high in this area. Area D is a heavily polluted area with some heavy industries, a big power plant (petroleum) and intensive use of fertilisers in agriculture. The maximum of area E coincides with an area where several glass manufacturing industries are located. (Compare with lead.) The high cadmium contents of area F and G are caused by industrial activities. Area H is under investigation. The bedrock in this area cannot be the cause of this anomaly. There are no known antropogenic activities in this area which could be the sources of these high cadmium values. There is however a cooperation going on between the environmental authorities and the geochemists in this matter. Medical authorities are also interested.

## Conclusions

It is clearly shown that the sample type used in our geochemical mapping (living roots and aquatic mosses) in a very sensitive way gives information on both the natural background and antropogenic sources. Research going on also shows that the organic sample type used is promising also for detecting other pollutants (*e.g.* PCB).

## References

Brundin, N.H., Ek, J.I. and Selinus, O. (1987). Biogeochemical studies of plants from stream banks in northern Sweden. *J. Geochem. Explor.*, 27, 157-188.
Selinus, O. (1983). Regression analysis applied to interpretation of geochemical data at the Geological Survey of Sweden. In: R.J. Howarth (ed.), *Handbook of Geochemistry* part 2; Statistics and Data analysis in geochemical prospecting. Elsevier, Amsterdam.

# 3 Applications of Exploratory Data Analysis to Regional Geochemical Mapping

C. Reimann, H. Kürzl and F. Wurzer
*Joanneum Research Society, Mineral Resources Research Division,*
*Roseggerstrasse. 15, A-8700 Leoben, Austria*

## Summary

*Methods of Exploratory Data Analysis (EDA) are used for quality control purposes, data analysis, documentation and mapping. In quality control these methods can greatly reduce the number of samples necessary to monitor accuracy and precision. Used for raw data documentation, EDA will often disclose unexpected data behaviour. For mapping purposes, EDA offers elegant methods for outlier definition and class selection that will uncover inherent data structures.*

## Introduction

The wide range of statistical methods used for interpreting geochemical data is conspicuous. Most of the methods are based on the assumption of (log-)normally distributed data. Rarely it is realised that these methods may give misleading and erroneous results if the data do not follow the supposition of (log-)normality.

Experience in handling large amounts of data from regional geochemical surveys in Austria showed that usually the data are not (log-)normal. Methods independent of any distributional model and solely reflecting the data structure are needed.

Exploratory Data Analysis (EDA) may be the solution to this problem. EDA incorporates a great variety of methods that give a better insight into data behaviour prior to the treatment of data by any probabilistic model. Resistant methods, as used in EDA, were proven to be extremely valuable tools when dealing with "soft" geochemical data.

21

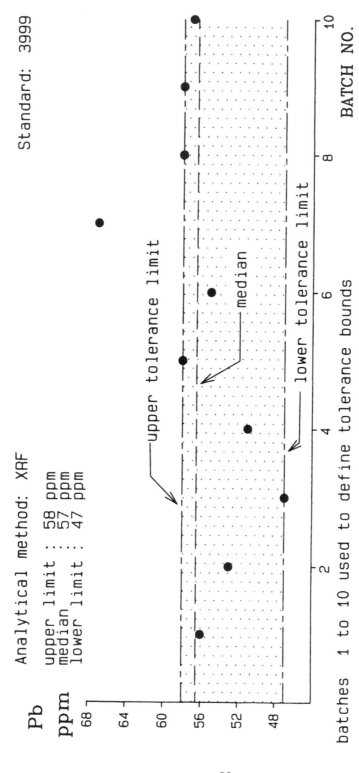

**Figure 1.** *Analytical results of a CRS. Tolerance limits are defined using the boxplot. Bias (median shifted towards upper tolerance limit) and 1 outlier could be recognised with as few as 10 determinations of the CRS.*

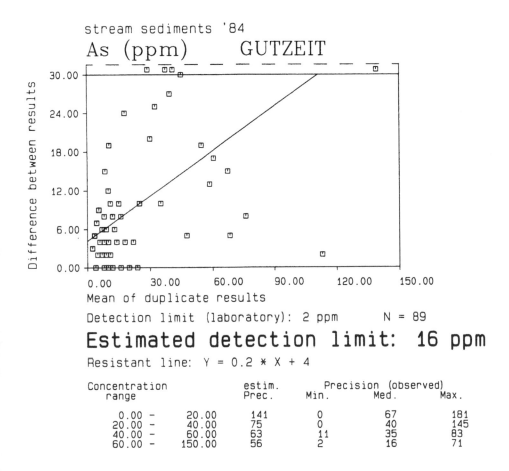

stream sediments '84

As (ppm)     GUTZEIT

Detection limit (laboratory): 2 ppm     N = 89

**Estimated detection limit: 16 ppm**

Resistant line: Y = 0.2 * X + 4

| Concentration range | | estim. Prec. | Precision (observed) | | |
|---|---|---|---|---|---|
| | | | Min. | Med. | Max. |
| 0.00 – | 20.00 | 141 | 0 | 67 | 181 |
| 20.00 – | 40.00 | 75 | 0 | 40 | 145 |
| 40.00 – | 60.00 | 63 | 11 | 35 | 83 |
| 60.00 – | 150.00 | 56 | 2 | 16 | 71 |

**Figure 2.** *Determination of the practical detection limit and precision via blind duplicates. If precision is as bad as in this example only very few classes should be mapped.*

**Quality Control (QC)**

QC is generally based on the insertion of a control reference sample (CRS) and blind duplicates (BD) of project samples at a rate of 1 each in every batch of 20 samples.

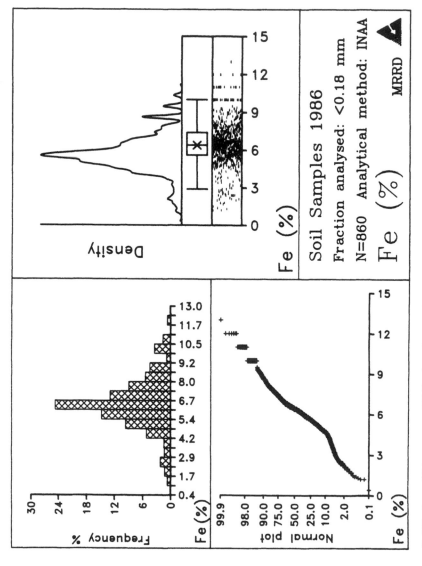

**Figure 3.** *Display of histogram, CDF-plot, density trace, boxplot and scattergram for Fe-analysis on soil samples. The density trace reveals the presence of several populations hidden in the histogram. The scattergram reveals discretisations for analytical values higher than 9.5% due to rounding at the laboratory.*

The aim of QC-procedures is to guarantee reliable and comparable analytical data for a geochemical survey. Data integration and adjustments at later times will only be possible when QC becomes an important part of any geochemical survey. As mapping of geochemical variability is the aim of a geochemical survey, precision must be known in order to select the appropriate number of class intervals. QC will supply this information.

*Control Reference Sample (CRS)*
Tolerance bounds for the CRS are normally calculated after 20-30 determinations of the CRS using the mean ± 2.5 standard deviations.

Using the boxplot (Tukey, 1977) to establish tolerance bounds for the CRS (Figure 1) has the following advantages:
- Tolerance bounds can be defined after as few as five determinations.
- Any bias of the analytical method will be recognised immediately.

*Blind Duplicates (BD)*
BDs are used to calculate analytical precision for any pair of BD-analyses. The results are then compared with the precision demanded from the laboratory by contract at the beginning of the project.

In addition, precision and practical detection limits can be estimated for the whole set of BD-analyses by using the method of Thompson and Howarth (1978). The introduction of resistant line methods (Figure 2) to this procedure has some major advantages (Reimann and Wurzer, 1987):
- the method will work reliably with as few as ten duplicate analyses (instead of more than 50 required for the commonly-used method);
- the possibility of fitting resistant lines according to piecewise linear behaviour;
- the estimation of the practical detection limit will not be influenced by changing precision at high concentrations.

To judge the sampling error a variance analysis based on additional replicate field samples may be required.

## Data Analysis and Documentation

EDA offers numerous graphical methods for data analysis and documentation. For interpretation of single variables a combination of histogram, CDF-Plot, density trace, one- dimensional scattergram and boxplot will give an excellent first insight into data structures (Figure 3) (Kürzl, 1987).

The one-dimensional scattergram is a very simple graphical display but gives more definite information on data behaviour (*e.g.* discretisations, gaps in the data structure, number of outliers) than the histogram.

The conventional histogram is very well suited to hide data structures but its popularity still persists in the literature. When compared with the density trace, it is easy to judge the amount of useful information hidden behind histogram bars. Thus, with ready access to computers the histogram should be replaced by the density trace.

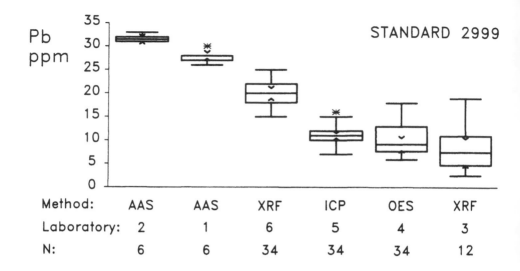

**Figure 4.** *Boxplot comparison of the analytical results of 6 different laboratories on a stream sediment standard. This may give an idea of how close the analytical results of the method/laboratory used for the survey are to the "true" element contents of the samples. In addition, a direct visual comparison of the quality of different methods/laboratories is possible.*

The boxplot will show the following features of a data batch at a glance: location, spread, skewness, tail length and outlying data points. This makes the boxplot extremely well suited to compare data batches (Figure 4) (Tukey, 1977; Hoaglin *et al.*, 1983).

**Mapping**

The selection of classes for (colour-)coding and the definition of outliers are crucial steps in the production of reliable geochemical maps. A subjective procedure, using arbitrarily chosen class boundaries (*e.g.* 5, 10, 20 or 50 ppm steps) will have no relation to the natural process and should thus be avoided. Instead, an empirical method based on data structures is needed.

The boxplot offers such an objective technique for automated routine class selections and outlier definition. The boxplot provides a maximum of 4 classes to map the body of data minus the outliers. Two optional classes for upper and lower outliers are only used when there are outliers present (Figure 5).

If the distribution of the analytical values to be mapped is extremely skewed, less than 4 classes will be used automatically (Figure 6). A maximum of 6 classes for routine class selections will avoid mapping process errors of up to 16%.

In summary the most important advantages of boxplot class selections and outlier definition include:

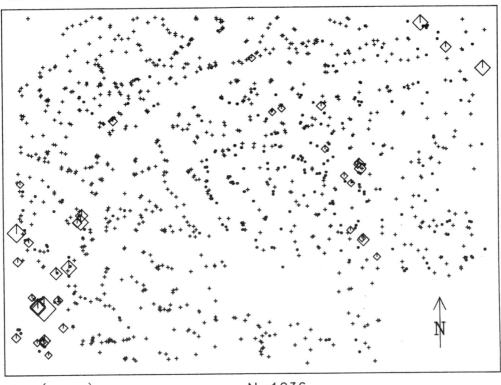

As (ppm)        N=1036

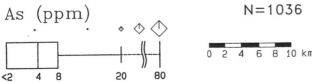

Figure 6. *Map of the As-distribution in stream sediments (size of area: 1,000 km². Analytical Method: Gutzeit (compare Figure 2). Mapping many classes will result in a map showing analytical variability. Using the boxplot, for this data set, only three classes are defined automatically. This is the absolute maximum that should be mapped if the map is to be reliable. The size of the outlier-symbol grows proportionally to the analytical value. Considering the practical detection limit it might be even better to map only two classes: "background" and "outliers".*

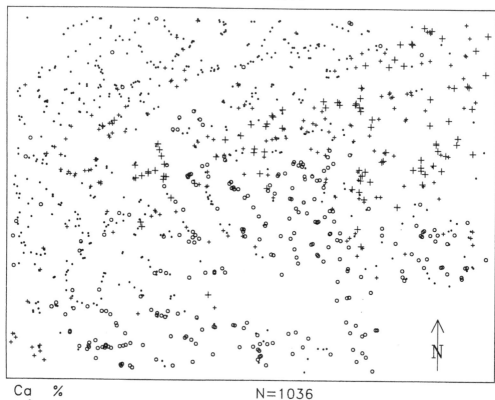

Ca   %                                    N=1036

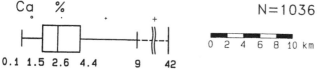

0.1 1.5 2.6  4.4          9      42          0  2  4  6  8 10 km

**Figure 5.** *Map of the Ca-distribution in stream sediments in an area of 1000 km². Class boundaries are defined by the boxplot. For mapping, EDA-symbols (Vellemann and Hoaglin, 1981) are used. These have the advantage that they give a balanced visual impression. Areas with low values can easily be distinguished from areas with medium and/or high Ca-concentrations. As an additional advantage areas with a high element variability can be recognised at once. Note that no lower outliers are present and the two possible classes within the box are presented by only 1 EDA-symbol.*

- up to 25% of "wild" data can be accommodated without influencing class boundary selection;
- outliers are objectively recognised;
- the actual data structure is mapped;
- mapping of "error" can be avoided;
- it can be used for automated, routine map production;
- the resulting maps can be used for many purposes (and not only exploration).

Experience has shown that class selections by the boxplot very often result in a map having a direct relation to the natural process, *e.g.* picking up the regional geology.

### References

Hoaglin, D.C., Mosteller, F. and Tukey, J.W. (1983). *Understanding Robust and Exploratory Data Analysis*. Wiley, New York.

Kürzl, H. (1988). *J. Geochem. Explor.*, **30**, 303-322.

Reimann, C. and Wurzer, F. (1986). *Mikrochimica Acta*, (Wein) 1986 II, 31-42.

Thompson, M. and Howarth, S.R. (1978). *J. Geochem. Explor.*, 9, 23-30.

Tukey, J.W. (1977). *Exploratory Data Analysis*. Addison Wesley, Reading, Massachusetts

Vellemann, P.F. and Hoaglin, D.C. (1981). *Application, Basics and Computing of Exploratory Data Analysis*. Duxbury Press, Boston, Massachusetts.

# 4 Baseline Trace Metal Survey of Welsh Soils with Special Reference to Lead

B.E. Davies and C.F. Paveley
*School of Environmental Science, University of Bradford, Bradford, West Yorkshire, BD7 1DP, England*

## Summary

*An account is given of trace metals, especially lead, in surface soils and subsoils in Wales. Samples were taken on a regular grid at 5 km intervals. Background levels for soil metals are estimated. Lead contamination of surface soils is widespread and can be accounted for by industry and past lead mining activity.*

## Introduction

Most studies of trace metals in Welsh soils have concentrated on contaminated areas. Griffiths (1919) ascribed infertility problems in the Ystwyth valley of west Wales to residual accumulations of lead originating during the main mining period of the mid-nineteenth century. Subsequently, Alloway and Davies (1971) confirmed that soils in the Ystwyth and neighbouring valleys were contaminated by lead and they extended the results to other metals, especially cadmium, copper and zinc and to other parts of Wales where lead mining had been active. Davies and Roberts (1978) made a detailed survey of soils in north-east Wales and produced isoline (contour) computer maps of soil metals around Halkyn and thereby evaluated the extent of metal pollution.

These and other detailed studies drew attention to the role of water in spreading pollutants. This process was especially active during the heyday of mining when, prior to the 1876 Rivers Pollution Act, mine effluent was discharged direct to rivers and led to valley-bottom soils being covered by metal-rich silt during periods of flooding. Other agencies of pollution which have been identified include the loss of material from old waste piles and, in the past, fallout from smelters.

A number of other accounts of metals in Welsh soils have been published. Burton and John (1977) have described heavy metal contamination arising from industry in soils of the Rhondda Fawr. Bradley *et al.* (1978) reported the concentrations of 12 trace elements in soil samples from 250 profiles in south west Dyfed. Bradley (1980)

29

has described trace metal concentrations in soils around Llechryd, Dyfed. Archer (1963) described the trace metal content of 5 soil profiles from Snowdonia.

With the exception of the work of Bradley *et al.* (1978) none of these papers attempted to assess normal levels of trace metals in soils in a comprehensive manner. Davies (1983) applied a graphical method to previous published data for soil lead and concluded that agricultural soils contained an average of 40 mg Pb kg$^{-1}$ with an upper limit of 120 mg kg$^{-1}$. The work upon which this account is based was intended to remedy this deficiency by analysing soils from many parts of Wales in order to estimate background concentrations.

## Methodology

The initial sample collection was undertaken by officers of the Soil Survey of England and Wales. Pedological A and B horizon samples were collected from soil profiles sampled throughout Wales on a regular grid 5 km apart. The original sampling plan allowed for 870 pairs of soil samples; not all these soil samples were available.

As far as was practicable unsampled sites were visited in August 1985 and the surface soils sampled using methods comparable with those of the first survey. The final collection comprised 824 A horizon and 646 B horizon samples.

All soils were either air dried (ca.15°C) or dried in a laboratory microwave oven. They were then disaggregated and sieved through nylon sieves of 2 mm aperture. Subsamples were then extracted, for "total" metal content. Subsamples (5g) of soil (105°C dry) were first treated with 20% hydrogen peroxide to oxidise the humus and then extracted with hot (80°C) *Aqua regia* in a tall form conical beaker sealed with plastic "cling film". Soil digests were analysed by flame or flameless atomic absorption spectrometry for Pb, Zn, Cu, Cd, Ni, Co, Mn nd Fe. Analyses for pH (after 30 minutes equilibrium in 0.05M CaCl$_2$) and organic content (loss-on-ignition at 430°C) were also carried out.

In order to achieve acceptable quality control the following procedure was instituted. Surface soil (A) and subsoil (B) sample pairs were analysed in random order in order to avoid differences in results consequent on long-term analytical drift. The analytical work was divided into batches; each batch comprised 25 A horizon and the associated 25 B horizon samples. Within each batch, 20% of the samples were duplicated and these were also chosen at random.

Accuracy was checked by including duplicate samples of a local reference soil in each batch. In addition, samples of the National Bureau of Standards (NBS) of the United States of America River Sediment (SRM 1465) or the EEC Bureau of Certification and Reference (BCR) reference soils (Nos 141 and 144) were included in selected batches. Mean % recoveries were Pb = 103%, Zn = 85%, Cu = 90%, Cd= 98%, Co = 107%, Ni = 92% and Mn = 86%. A metal composite solution standard was also analysed in each batch to check instrument performance.

Precision is a measure of agreement between replicates and is usually quoted as the % Coefficient of Variation (CV) which is the standard deviation divided by the arithmetic mean and multiplied by 100. Generally, the CV for all metals was <5%.

30

## Results

### The Numerical Data

The analytical results for 824 A horizon and 646 B horizon soil samples are summarised in Table 1. Except for pH and Ni the median is seen in all cases to be smaller than the arithmetic mean thereby indicating positively skewed populations. For many statistical tests it is necessary to normalise skewed distributions and this was done by making a $\log_{10}$ transformation of the data. Table 1 includes the geometric mean and the geometric deviation of the transformed data (after antilogging) and it can be seen that the transformation has brought the geometric means and arithmetic medians closer together implying that the distribution of each variable is approximately log- normal.

Maxima and minima yield useful information concerning anomalous values but are not useful in establishing typical ranges. One approach is to estimate the percentiles of a frequency distribution of each element from a plot of cumulative percent lead frequency. This was done here after a log-transformation and the results for lead are reported in Table 2. Drawing and reading these curves involve approximations and therefore the 50th percentile values are not necessarily identical with the geometric means reported in Table 1.

From Table 2 it is seen that only 5% of topsoil Pb concentrations exceed 182 mg Pb kg$^{-1}$ soil. This puts in perspective literature reports of areas of severe Pb contamination: most of Wales contains soils containing <182 mg Pb kg$^{-1}$.

The mean Pb content of surface soils is 76 mg kg$^{-1}$ and the median is 36 mg Pb kg$^{-1}$ (Table 1). Davies (1983) calculated that the mean normal lead content of uncontaminated Ceredigion and Halkyn Mountain soils was 37 and 52 mg kg$^{-1}$, respectively. The grid coverage included soil samples from old lead mining areas where contamination is known to occur and these high values have had the effect of weighting the mean whereas the median is less affected and represents a better measure of the "average" soil lead content. The geometric mean is 42 mg kg$^{-1}$ and this too is in closer accord with previously published values (Davies, 1983) for lead in uncontaminated soils. Subsoil mean Pb (34 mg kg$^{-1}$, Table 1) is lower than surface soil Pb.

The maximum surface soil Pb value (3369 mg kg$^{-1}$) was for a soil sample derived from Halkyn Mountain where Davies and Roberts (1978) reported that 17% of their study area contained >1000 mg Pb kg$^{-1}$ soil. The maximum subsoil value (2096 mg Pb kg$^{-1}$) also derived from Halkyn Mountain.

The grid survey data have therefore highlighted the soils of Halkyn Mountain as being severely lead contaminated but "typical" soil Pb concentrations in Wales are comparable with those found previously in south west Wales or generally elsewhere.

Minimum soil Pb values are very low, 1.3 and 0.36 mg kg$^{-1}$ for A and B horizons respectively. The topsoil value derived from east of Llandeilo, Dyfed (grid 266216) and the subsoil from an immature sandy soil in Anglesey (grid 241366).

### Topsoil/Subsoil Ratios

The high rainfall over most of Wales implies that Welsh soils are likely to be well or strongly leached. Under these profile drainage conditions surface soils become

31

**Table 1.** *Summary statistics for 824 A Horizon and 646 B Horizon soil samples from profiles in Wales. Metal values are mg kg$^{-1}$ dry soil.*

| | | Min | Max | Mean | Median | Standard Deviation | Geometric Mean | Geometric Deviation |
|---|---|---|---|---|---|---|---|---|
| Pb | A | 1.3 | 3369 | 76 | 36 | 186 | 4 | 2.6 |
| | B | 0.36 | 2096 | 34 | 17 | 132 | 17 | 2.3 |
| Zn | A | 4.7 | 2119 | 83 | 63 | 107 | 60 | 2.2 |
| | B | 1.1 | 1451 | 67 | 58 | 76 | 51 | 2.2 |
| Cu | A | 0.13 | 215 | 17 | 12 | 20 | 12 | 2.3 |
| | B | 0.09 | 65 | 13 | 11 | 8.7 | 9.5 | 2.3 |
| Cd | A | 0.01 | 15 | 0.60 | 0.36 | 0.87 | 0.33 | 3.0 |
| | B | 0.01 | 12 | 0.36 | 0.12 | 0.83 | 0.14 | 4.4 |
| Co | A | 0.15 | 2284 | 13 | 8.0 | 100 | 6.6 | 2.5 |
| | B | 0.14 | 43 | 10 | 9.3 | 5.9 | 8.1 | 2.1 |
| Ni | A | 0.42 | 169 | 16 | 15 | 13 | 13 | 2.2 |
| | B | 0.63 | 80 | 20 | 19 | 12 | 16 | 2.2 |
| Mn | A | 3.0 | 496094 | 2266 | 599 | 24823 | 417 | 45 |
| | B | 1.7 | 10555 | 744 | 550 | 789 | 427 | 37 |
| Fe | A | 38 | 111505 | 16394 | 13657 | 12789 | 11220 | 295 |
| | B | 12 | 101125 | 16164 | 13649 | 10870 | 12303 | 251 |
| pH | A | 2.0 | 7.8 | 4.4 | | | 4.5 | 0.92 |
| | B | 1.8 | 7.9 | 4.7 | | | 4.8 | 0.93 |
| %OM | A | 0.29 | 99 | 11 | | | 5.5 | 15 |
| | B | 0.11 | 99 | 5.6 | | | 2.4 | 12 |

generally impoverished in trace elements and B horizons may become enriched through capture mechanisms. In contrast, cycling through vegetation, especially trees, may lead to some enrichment of surface soils in association with the higher amounts of humified organic matter found there. Earlier papers which have reported the tendency for some elements to be enriched in surface layers even in well leached soils failed to recognise the importance of anthropogenic inputs. Archer (1963) found no consistent pattern for profile distribution of trace elements in Snowdonian soils

**Table 2.** *Percentile values from the frequency distribution for soil lead concentration.*

| Percentile | Soil Lead (mg kg$^{-1}$) | |
| --- | --- | --- |
| | A horizon | B horizon |
| 5 | 0.8 | 7.0 |
| 10 | 12 | 7.4 |
| 20 | 19 | 11 |
| 30 | 23 | 15 |
| 40 | 26 | 16 |
| 50 | 29 | 18 |
| 60 | 35 | 20 |
| 70 | 44 | 22 |
| 80 | 66 | 28 |
| 90 | 115 | 40 |
| 95 | 182 | 68 |
| 99 | 407 | 525 |
| 99.9 | 631 | 2512 |

except that more Co could be extracted by dilute acetic acid from surface soils and there was a general surface accumulation of Pb. In Lancashire soils Butler (1954) reported that all soils show surface accumulations of Sn and Pb. In Scotland Swaine and Mitchell (1960) noted that the surface accumulation of Pb is the outstanding effect observed in all profiles otherwise surface soils tended to be depleted in trace elements relative to parent materials.

Soil science lacks any predictive theory for trace element leachability but there is a general consensus that surface soils are likely to be depleted in trace elements but this may be offset to some unknown extent by accumulation in surface humus- rich horizons. Indications of a marked accumulation of any trace element in soil profiles may therefore indicate contamination. This surface enhancement has been remarked upon in the discussion of the individual elements and is summarised in Table 3 where the ratios of the geometric mean topsoil/geometric mean subsoil concentrations are presented. Two elements, Cd and Pb, are seen to have ratios >2 suggesting marked surface accumulations and therefore widespread contamination. This is in accordance with the hypothesis that centuries of mining and smelting Pb and Zn ores has caused soil contamination in areas well away from individual smelters or mining areas. But the source of Cd is zinc ore and only a small enhancement of surface Zn is seen. Again, this raises the possibility that Zn is relatively readily leached from surface soils whereas Cd may be retained in A horizons. This may have profound environmental consequences in view of the toxicity of Cd and its ease of uptake by crop plants. Copper also displays some surface enhancement. In contrast, the four

**Table 3.** *Topsoil/Subsoil Ratios for Welsh Soils (Ratios of geometric mean concentrations, in mg kg$^{-1}$).*

| Pb | Zn | Cu | Cd | Co | Ni | Mn | Fe |
|------|------|------|------|------|------|------|------|
| 2.5 | 1.2 | 1.3 | 2.5 | 0.82 | 0.81 | 0.98 | 0.91 |

elements Co, Ni, Mn and Fe, which are known to be especially prone to leaching in podzolic soils show surface depletions.

### Maps of Soil Metals

The next stage of the analysis was the preparation of plots of metals in Welsh soils. The data were categorised for plotting using the percentiles enumerated in Table 2. The percentiles chosen were the 70, 90, 95 and 99.

The plot for topsoil Pb (Figure 1) confirms the contaminated environment located on Halkyn Mountain. Sporadic highs are observed in topsoils throughout mid-Wales where there are numerous, scattered lead mines but the Cardiganshire ore field is not prominent. Here, the contaminated soils are often alluvial soils in the form of long strips running up valleys and grid surveys often under-represent such long, thin areas. High values in the Vale of Glamorgan, west of Bridgend are curious. There were lead mines there, in the Carboniferous limestone, but coal mining superseded lead mining throughout south Wales and the lead workings were usually abandoned. High values are also located north and east of Swansea and along the heads of the coalfield valleys and are presumed to be the consequence of smelting and general industrial activity. A small area of high lead values at Milford Haven may be the consequence of pollution from a coal- fired power station at Pembroke Dock.

### Conclusions

This survey has provided both statistical and cartographical information for trace metals in Welsh soils. The comprehensive nature of the survey has allowed the estimation of background concentrations and has provided these values for both surface soils and subsoils. Previous reports by several authors have concentrated on known contaminated areas which have yielded very high soil metal contents. This survey confirms that such high metal concentrations do occur but only 5% of Welsh surface soils exceed 182 mg Pb kg$^{-1}$. The computergraphic plots have shown where in Wales high values may be found and they confirm that lead contamination is widespread in the Halkyn Mountain mining area of north east Clwyd and in industrial south Wales. Contaminated areas occur sporadically throughout the rest of Wales and can be associated with old lead mine workings.

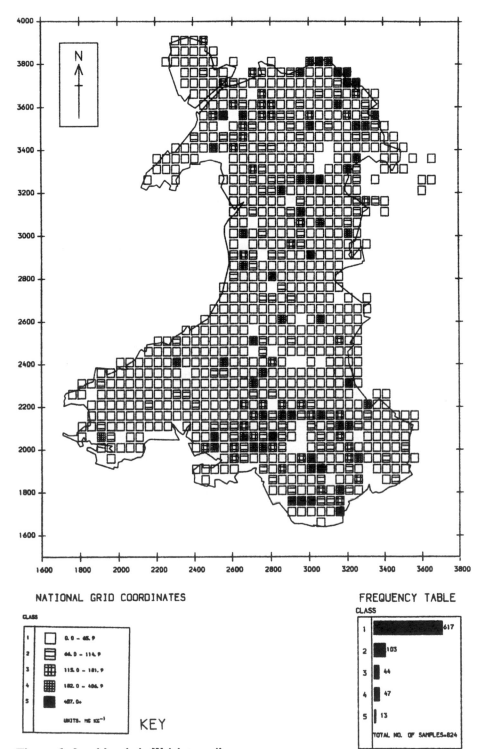

**Figure 1**. *Lead levels in Welsh topsoils.*

NATIONAL GRID COORDINATES

KEY

CLASS

| | | |
|---|---|---|
| 1 | | 0.0 – 45.9 |
| 2 | | 46.0 – 114.9 |
| 3 | | 115.0 – 181.9 |
| 4 | | 182.0 – 406.9 |
| 5 | | 407.0+ |

UNITS. MG KG$^{-1}$

FREQUENCY TABLE

CLASS

| | |
|---|---|
| 1 | 617 |
| 2 | 103 |
| 3 | 44 |
| 4 | 47 |
| 5 | 13 |

TOTAL NO. OF SAMPLES=824

## Acknowledgement

This work was commissioned and financed by the Welsh Office, Cardiff, and the authors record their gratitude for the support given.

## References

Alloway, B.J. and Davies, B.E. (1971). Trace element content of soils affected by base metal mining in Wales. *Geoderma*, **5**, 197- 208.

Archer, F.C. (1963). Trace elements in some Welsh upland soils. *J. Soil Sci.*, **14**, 144-148.

Bradley, R.I., Rudeforth, C.C. and Wilkins, C. (1978). Distribution of some chemical elements in the soils of north west Pembrokeshire. *J. Soil Sci.*, **29**, 258-270.

Bradley, R.I. (1980). Trace elements in soils around Llechryd, Dyfed, Wales. *Geoderma*, **24**, 17-23.

Burton, K.W. and John, E. (1977). A study of heavy metal contamination in the Rhondda Fawr, South Wales. *Water, Air and Soil Pollution*, **7**, 45-68.

Butler, J.R. (1954). Trace element distribution in some Lancashire soils. *J. Soil Sci.*, **5**, 156-166.

Davies, B.E. and Roberts, L.J. (1978). The distribution of heavy metal contaminated soils in north east Clwyd, Wales. *Water, Air and Soil Pollution*, **9**, 507, 518.

Davies, B.E. (1983). A graphical estimation of the normal lead content of some British soils. *Geoderma*, **29**, 67-75.

Griffith, J.J. (1919). Influence of mines upon land and livestock in Cardiganshire. *J. Agric. Sci.*, **IX**, 366-398.

Swaine, D.J. and Mitchell, R.L. (1960). Trace element distribution in soil profiles. *J. Soil Sci.*, **11**, 347-368.

# 5 A Survey of Chemometric Techniques for Exploratory Data Analysis and Pattern Recognition

Robert R. Meglen

*Center for Environmental Sciences, Laboratory for Chemometrics, 1100 14th St., University of Colorado at Denver, Denver, CO 80202, USA*

## Introduction

Research in the area of environmental geochemistry and health challenges chemists, geologists and medical scientists as they search for the links between the geosphere and biosphere. The diversity of scientific disciplines represented in this conference reflects the complexity of the system under study. Environmental research requires experimental designs that are capable of generating data on hydrological, mineralogical, biological and chemical processes. If one examines the recent history of research in environmental geochemistry and health it is apparent that data on a large number of variables is usually acquired. However, massive quantities of data and collaboration alone will not ensure that an adequate mechanistic description of the system will be obtained. The problem is that in complex environmental systems significant patterns in the data are not always obvious when one examines the data by conventional data analysis methods, one variable at a time. For data interpretation purposes, complexity may be viewed as interactions (correlations or covariances) among many variables. Interactions among the measured variables tend to dominate the data in complex systems and this useful information is not extracted by univariate approaches. The existence of many variables from each of several measurement domains (microbiology, chemistry, geology, hydrology, *etc.*) suggests that multivariate statistical techniques should be used to examine the complex data bases.

Most powerful mathematical and statistical techniques in the physical sciences have a sufficient historical record of application that justifies including them in one's computational repertoire. Conventional statistical procedures such as linear regression and computation of means and dispersion indices are so widely used that they are resident options of most pocket calculators. The long history of using multivariate statistics in the social sciences has made their use commonplace in those fields today. Indeed most social science curricula mandate formal training in multivariate statistics. However, these techniques have not been widely used in the physical sciences. The reasons for this are not clear; but disciplinary prejudice and disparate philosophies have contributed to the slow acceptance of these techniques in physical sciences. Most physical scientists are unfamiliar with the literature of the social sciences and multivariate techniques. However, during the past decade an increasing number of chemists have "rediscovered" the power of multivariate statistical methods and have begun exploring their utility in chemical problem solving. Thus, chemometrics, a new chemical subdiscipline has been developed.

**Chemometrics** is the *chemical* discipline that uses mathematical and statistical methods

1. to design or select optimal measurement procedures and experiments, and
2. to provide maximum chemical information by analysing chemical data.

It is important to recognise why chemists have found it necessary to begin a new field of study. Recent advances in instrumental design, microprocessor control and computer acquisition of data have increased the rate at which data are obtained. It is now possible to obtain large numbers of measurements on many parameters in a fraction of the time previously required for much smaller efforts. Through automation we are making both qualitative and quantitative changes in the way we perform experiments. The fundamental philosophy of experimentation and its practical expression through experimental design have drastically changed. Unfortunately our interpretive abilities have not improved at a rate that permits full exploitation of this electronic windfall. The added mass of data is at best under-used and at worst tends to obscure the hidden information rather than clarify it. Full exploitation of our newly acquired data affluence requires enhanced interpretive aides.

The field of chemometrics has developed to facilitate the transfer of statistical techniques to the chemical problems that require them. We emphasise that chemometrics is a chemical discipline because the interpretive power of the statistical techniques can be fully exploited only when a chemical context is present.

We begin by emphasising that data are not information! As indicated earlier, data are easy to collect. Any machine can be programmed to mindlessly acquire numbers. What one really desires is **information**. One goal of chemometrics is to provide tools (statistical and mathematical methods) to assist the conversion of data to information. The central focus of this discussion involves the conversion of data into information and, ultimately, into knowledge that assists the problem solving endeavour.

### The Rationale for Using Pattern Recognition

One approach to assisting in the conversion of data into information is called pattern recognition (Harper *et al.*, 1977; Kowalski, 1977; Erickson *et al.*, 1980; Massart *et al.*, 1980). The rationale for using pattern recognition to examine large multivariate data bases is to enhance human understanding of the multidimensional information contained in the data. The mathematical techniques employed in pattern recognition permit rapid and efficient identification of relationships and key aspects that otherwise might remain hidden in the large mass of numbers.

Recent advances in electronics and automated data collection techniques make it possible to perform simultaneous measurements of dozens of variables on hundreds of samples in a fraction of the time it previously took to accomplish much smaller tasks. Data tables containing large numbers of measurements and samples yield slowly to traditional data analysis because these techniques are limited to handling one or two variables at a time. To expedite the analysis and enhance the understanding of complex problems it is necessary to have either prior human experience on similar problems, or data analysis techniques that can rapidly extract the important variables and identify the underlying relationships that distinguish among the samples.

Without prior experience or computer assisted pattern recognition the evaluation of a 30 variable data matrix would be a formidable task. Since it is usually of interest to determine whether there are any significant bivariate relationships (correlations) among the measured variables, one would need to examine a total of 435 (30 x 29/2) scatter plots. If one were to seek discrimination among selected sample groups additional plots would be required. Thus, complete examination of the data base could require preparing and examining hundreds of unique plots. It is likely that only a few of these plots would yield insight about the relationship of the measured variables. Using this "brute force" interpretation technique could still miss relationships that involve more than two variables at a time. In addition to the quantitative difficulties attending the data base examination there are several qualitative limitations imposed by the nature of the measurements themselves. Many standard statistical techniques require some knowledge or assumptions about the shape of the distribution of measured values. Measurements on natural phenomena often exhibit poorly behaved distributions; some are highly skewed and some are multimodal. Robust analysis of these data requires techniques that do not rely upon a prior knowledge or assumptions about the underlying variable distributions. Qualitative limitations and the magnitude of the interpretive task mandates a computer assisted examination of all possible relationships.

## Pattern Recognition Principles

There are several advantages that obtain from the application of pattern recognition for data interpretation. The methods can rapidly identify key variables that are important in making distinctions among samples. This greatly reduces the number of two dimensional plots that must be examined since the technique extracts the relatively few plots that are likely to be most effective in displaying differences that make a difference. Another advantage of these techniques is that they are multivariate, they incorporate many variables simultaneously. As a result they can indicate subtle changes that slightly affect several variables simultaneously. This type of change is easily missed when examining only one or two variables at a time. In addition, the technique gives equal weight to variables with small absolute values relative to other large variables. This feature ensures that the search for relevant variance among small numbers is not obscured by large invariant features.

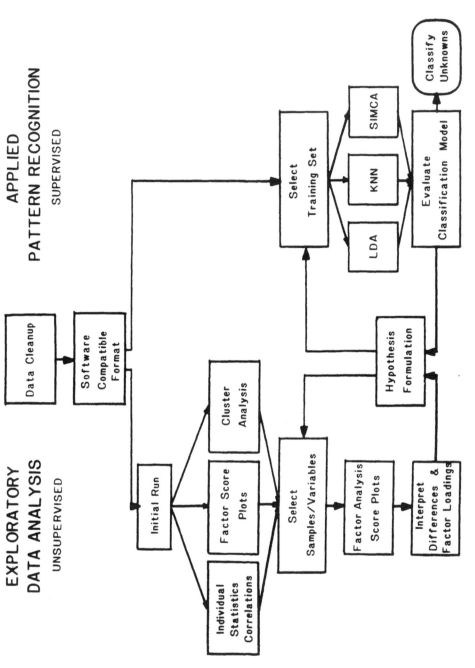

**Figure 1.** Schematic diagram illustrating the stepwise iterative processes used in exploratory data analysis and applied pattern recognition.

40

The pattern recognition approach consists of two phases; exploratory data analysis and applied pattern recognition (classification model development). The purpose of the exploratory phase is to uncover the basic relationships that exist among the variables and between the samples. The applied pattern recognition phase tests the strength of these basic relationships and other presumed relationships by developing classification/prediction models and determining their accuracy. A brief description of both phases is presented below, but the simple flow diagram shown in Figure 1 provides an outline of the process.

The diagram shows that before any work can proceed it is necessary to prepare the existing data base for computerised examination. A data clean-up step consists of a search for several potential problems that could restrict the full utilisation of the existing data. Columns of measurements that are incomplete can be "filled" if only a small percentage of samples have missing measurements. It is possible to "fill" the missing data items in an unbiased way so that all of the measurements can be used. If most of the measurements are missing one must delete the variable from further consideration. Chemical data are often entered as "below detection limit". Designation as below detection limit is a quantitative determination, *i.e.* it contains useful information relative to all samples that exceed the detection limit. There are a variety of protocols that permit us to pre-treat detection limit data and enter them into the data base during the clean-up phase. Once the data base has been prepared for use it is placed into a storage format for the computer.

Exploratory data analysis is designed to uncover three main aspects of the data: anomalous samples or measurements, significant relationships among the measured variables, significant relationships or groupings among the samples. Exploratory data analysis is an iterative process in which a wide variety of tools are employed. There is no set sequence in which these tools are applied. Each data base may be approached in a different way, but after all of the iterations and alternate paths have been explored the key findings should converge to a single coherent summary of the data base. One commonly employs two different approaches when examining large data bases. The first approach is consistent with the most basic assumption of the exploratory analysis, that all of the data are "good" and that nothing is known about the structure of the data base. This approach is particularly useful when other interpretation attempts by techniques other than pattern recognition have been exhausted. This approach is powerful since it does not impose a bias about the data base that precludes exploring so-called unfruitful paths. By initially including all of the data, regardless of any predisposition toward its value, we rely upon the pattern recognition algorithms to identify unusual behaviour. There is a fundamental philosophical reason for preferring this approach. Instead of searching the data for an answer, we ask the more fundamental question, "What do the data tell us?". By examining anomalous behaviour our attention is focused on distinctions that the data warrant and relationships that can be identified. Thus, in each iteration we identify a difference and then attempt to explain it. After successive layers of explainable results (information) have been peeled away only "noise" remains. As each anomaly is identified and confirmed by independently established knowledge of the data base, one gains insight about the system and confidence that useful information is being

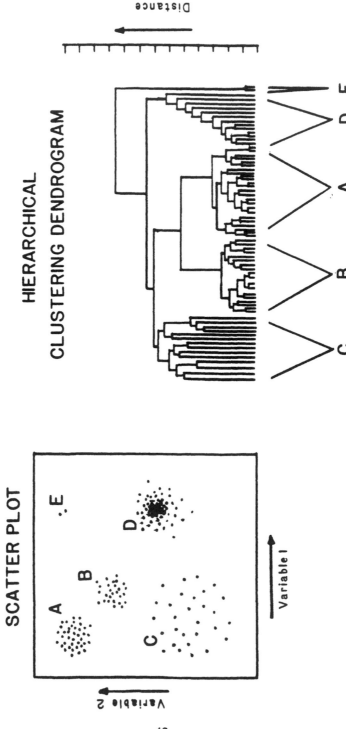

**Figure 2.** *Illustration of how two dimensional mapping clustering dendrograms may be used to identify class differences among objects.*

uncovered. The older literature on artificial intelligence calls this approach unsupervised learning. The three primary tools used in this approach are factor analysis, principal component analysis, and cluster analysis.

The second approach embodies a closer reliance upon prior knowledge of the system. One begins by accepting the status quo and searching for confirmatory evidence that it is correct. This is the applied pattern recognition procedure, sometimes called supervised learning. One begins with a set of samples whose class membership is well known; this constitutes a training set. Measurements on the known samples in the training set are used to construct a model that accurately represents the characteristics of this class. Samples which one wishes to classify into the proposed scheme (the test set) are compared to the classification model and assigned to the appropriate class. When anomalies not consistent with present knowledge are found new hypotheses and explanations are formulated. This approach employs the same basic tools of factor analysis and cluster analysis, but it relies more heavily upon classification modelling techniques such as linear discriminant analysis, nearest neighbour classification, and soft independent modelling by class analogy (SIMCA).

As indicated earlier we shall see that both approaches converge to a common interpretation of the data base and produce a self-consistent description of the system under scrutiny.

*Factor Analysis*

Factor analysis (Cooley and Lohnes, 1971; Davis, 1973; Gorsuch, 1974; Malinowski and Howery, 1980) typically consists of two steps: a strictly mathematical step called principal component analysis, followed by a refinement step that employs mathematical tools to enhance the interpretability of the extracted factors. The aim of factor analysis is to identify the few important dimensions (*i.e.* factors or "types" of variables) that are sufficient to explain the meaningful information in the data set.

The scatter plot shown in Figure 2 illustrates how one may detect significant associations among the samples on which only two variables have been measured. While the computed correlation among all of the plotted points is close to zero, there appears to be some association or clustering of points into well separated groups. Cluster A and B appear to be relatively dense but uniformly dispersed. They are qualitatively different from cluster C in which individual samples are also uniformly dispersed but separated by larger distances. Cluster D exhibits points that are not uniformly distributed over the cluster. There is a dense packing near the centre which gets more diffuse near the outer reaches of the cluster. Cluster E consists of only two points which appear to be quite similar to each other but very different (distant) from all other samples. It is clear that if one were to have constructed a one dimensional plot of these points on either one of the two axes the separate clusters would not have been discernible. Simultaneous representation of the data in two dimensions provides information not apparent in lower dimensional representations. This conclusion may be generalised to higher dimensional data bases. The fact that we cannot, by ordinary graphical techniques, construct simple visual representations for the multi-dimensional cases forms the basis of the factor analytic approach to multi-dimensional data analysis.

Since each measured parameter adds a dimension to the data representation, measurement of 30 variables requires the ability to depict relationships in a 30-dimensional space. This is well beyond the two or three dimensions which humans conceptualise comfortably. It is also beyond the graphical representation capabilities commonly used. Factor analysis is one of the pattern recognition techniques that uses all of the measured variables (features) to examine the interrelationships in the data. It accomplishes dimension reduction by minimising minor variations so that major variations may be summarised. Thus, the maximum information from the original variables is included in a few *derived* variables or *factors*. Once the dimensionality of the problem has been reduced it is possible to depict the data in a few selected two or three dimensional plots. We shall see how these plots highlight the significant features of the underlying data structure.

In addition to the graphical representations we also obtain a set of simple linear combinations of variables that enable us to quantify similar or parallel behaviours among the measured variables. These variable groupings permit us to generalise the behaviours into factors. Qualitatively different areas where little generalisation can be made between two areas are referred to as separate factors. An example of a factor might include a group of chemical elements which, upon inspection, suggests that the variables included in the factor characterise a particular mineral. Recall that these factors arise out of the natural association of these elements with one another, information derived from the chemical analyses recorded in the data, not from any structure imposed by the data analyst. The natural associations among the variables is quantified by computing the correlation coefficients among all variable pairs. The technique known as principal component analysis is accomplished by the mathematical tool of eigenanalysis. Eigenanalysis extracts the best, mutually independent axes (dimensions) that describe the data set. These axes are the so-called *factors* or *principal components*. They are linear combinations of the original variables. The utility of constructing a new set of axes to describe the data is that most of the total variance (information) in the data set may be concentrated into a few derived variables. This means that instead of having to depict the data on dozens of bivariate plots prepared from the original sample measurements, we can compute the location or *factor score* of each of the samples in the new data space. Thus we may depict most of the information on just a few two dimensional *factor score plots*. This process may be viewed as projecting the original data from its multidimensional representation down to two dimensions. As with any projection, information is lost; but this technique maximises the retention of information and quantifies the amount of information contained within each projection. In most chemical systems it is possible to depict 80- 90% of the total information in less than half a dozen plots. The two dimensional plots derived through principal component extraction on a multivariate data set would look very similar to the scatter plot shown in Figure 2. The only difference is that the axes are now linear combinations of all of the variables instead of single variables. Performing this mathematical transformation has permitted us to exploit the interpretive power inherent in graphical display techniques.

The second step in factor analysis is interpretation of the principal components or factors. This process is analogous to the interpretive step that may have resulted from our observation of separate clusters or sample groups that we illustrated in the two dimensional example given earlier. In the two variable example one might have concluded that separations between groups A, B and D were characterised by some intrinsic mechanism that causes variable 2 to decrease between groups while variable 1 increases. Similarly one might conclude that the distinction between sample group A and C is apparently due to changes in variable 2 and that it is unrelated to the sample's magnitude of variable 1. Our ability to identify different sample groups has led to enhanced insight with regard to the nature of the differences. Our ability to quantitatively associate these inter-group separations with "mechanisms" is derived from our knowledge of what the axes are depicting. In the multidimensional analog this is accomplished by examining the contribution that each of the original measured variables makes to the linear combination describing the factor axis. These contributions are called the *factor loadings*. When several variables have large loadings on a factor they may be identified as being associated. From this association one may infer chemical or physical interactions that may then be interpreted in a mechanistic sense. A small loading of a variable on a factor indicates that the variable is not associated with the other variables that comprise this factor; and that it is unimportant in making distinctions along this dimension. For research in the area of geochemistry and health the discovery of the variable associations provides insight about how otherwise obscure relationships express themselves in very different types of measurements. Conventional data analysis may reveal the chemical interactions to the geochemist. However, the added interpretive power provided by this approach is derived from the ability to include variables from many measurement domains. Thus, key interactions or associations between chemical, geological, biological variables are uncovered and presented for interpretation.

*Cluster Analysis*

Cluster analysis (Anderberg, 1973; Davis, 1973) uses techniques that search for unbiased natural groupings in the data, either among the samples or among the variables. Cluster analysis among the measurements can help to identify relationships among the variables. This application can provide additional confirmation of the factor loadings obtained in the factor analysis. Cluster analysis on the samples can identify natural groupings among the samples. These results may be compared to the results obtained from the factor plots described above. In general, the results of factor analysis and cluster analysis are complementary. When used in tandem they provide a powerful tool for the insight enhancement necessary to construct classification models in the final step of applied pattern recognition.

There are many clustering techniques that may be employed in data interpretation. For brevity we shall illustrate only one here. One type of hierarchical clustering is accomplished by quantifying similarities between samples by computing geometric distances between the samples in the variable space (in our example case, 30-dimensional space). Two samples that are very similar in all their measured variables would appear to be very close to one another if we could view the multidimensional representation in the complete data space. While this is difficult

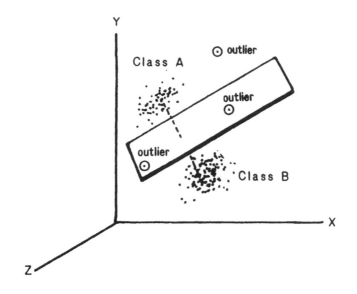

## Linear Discriminant Analysis
## LDA

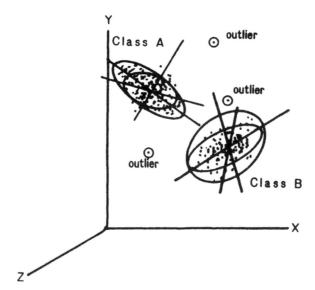

## Soft Independent Modeling by Class Analogy
## SIMCA

**Figure 3**. *Graphical illustration of two common classification methods. Linear Discriminant Analysis and Soft Independent Modelling by Class Analogy.*

for the human mind to comprehend, the computer can easily compute all of the inter-sample distances and compare them. If several samples lie very close to one another, but lie more distant from all other individual samples they constitute a cluster. The results of cluster analysis are usually depicted graphically in a diagram called a clustering dendrogram. A dendrogram is a two dimensional representation of the multidimensional relationships among all of the samples. An example of a hierarchical clustering dendrogram is shown on the right hand side of Figure 2. It shows how a cluster diagram constructed from the two variable example would appear. Each sample is represented by a point along the bottom of the plot. The two closest (most similar points in the data space) samples are joined (clustered) by a horizontal line drawn at their computed level of similarity (geometric distance in hyperspace). This is depicted along the vertical axis of the plot. The clustered samples are then treated as a single sample at their average position. The distance between all single samples and the new cluster are then compared. The two next closest samples are then clustered. If the previous cluster is closer to another individual sample than any other single sample, the new member joins the previous cluster. Their average position is used for successive comparisons. The process is repeated until all individuals have been assigned to a cluster. By examining the dendrogram it is possible to identify samples that are joined together at a high level of similarity. These constitute clusters that are naturally associated. Notice that the uniformly distributed groups (A and B) on the scatter plot appear as separate clusters on the dendrogram. The closeness of individual samples within the clusters may be discerned from the short vertical line segments. The uniformity of spacing between samples is also apparent from the relative similarity of vertical line segment lengths. The uniformly dispersed but wider separations within cluster C are evident in the dendrogram. Cluster D is characterised by a stair-step appearance along the distance axis of the dendrogram. The horizontal connection between clusters A and B shows that group A is closer to group B than any other group. The centroids of groups C and D are very nearly equidistant from the A and B pair. The two outliers (cluster E) are very close to each other, but quite distant from all other groups. While both clustering and factor analysis provide similar conclusions, as they should, the factor analysis plots are more insightful in deducing the reasons for class distinctions among the samples because the identity of the axes that produce separations is available. The clustering dendrogram only shows which samples are similar but not which variables are responsible for the similarity. This type of data analysis provides an unbiased approach to finding out how many fundamentally different groups are distinguishable by the measurements that were made.

## Classification Analysis Techniques

There are many algorithms that can be used in the classification of objects. Three of the most commonly used are : Soft Independent Modelling by Class Analogy (SIMCA), K-Nearest Neighbours (KNN), and Linear Discriminant Analysis (LDA). A complete description of all the techniques would require too much space for the

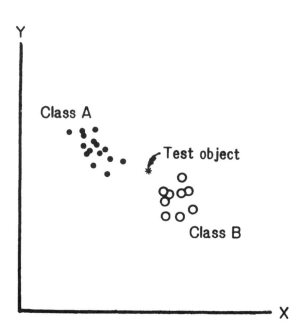

**Figure 4.** *Graphical illustration of the non-parametric classification by the K nearest neighbour methods.*

format of this presentation. However, a brief conceptual picture of SIMCA and KNN will be given.

*SIMCA*

SIMCA (Albano *et al.*, 1978) is a supervised learning algorithm that uses principal component analysis (factor analysis) to construct a separate principal component model to describe each of the classes hypothesised in the training set. The model is constructed from the category means plus as many additional principal components as necessary to describe the variance about the mean. Figure 3 illustrates two classes of objects whose means in three-space are well separated. In this example it is likely that the means alone would be sufficient to correctly classify most of the objects. However, for illustration purposes the axes of the first three principal components have been diagrammed. When the data are less symmetrically distributed about the mean, the principal component terms "adjust" the model so that the asymmetric spread is incorporated. If a class requires only one principal component term plus the mean we may conclude that the objects in this class are distributed along a line. A class requiring two principal components plus the mean would appear as a planar distribution. A mean plus three principal component class would be a solid (ideally an ellipsoid).

For example, if four separate classes of samples exist, one performs separate principal component analyses on each of the four samples groups. Each set of principal components is the best set of axes that describes the samples within the class. They permit one to "draw" a multidimensional surface which encloses all members in the class. When one wishes to classify a new object into one of the four hypothesised classes, one merely computes the location of the test object relative to each of the four sets of axes. If the object lies within a class enclosure it is assigned membership of that class. If it does not lie within the enclosure of any of the four classes, it is flagged as an "outlier". This modelling strategy may be called "exclusive" in that each test object is only assigned to membership of a class if it lies within the appropriate statistical boundaries. This feature is extremely powerful in classification analysis since it preserves the option that outliers may belong to an unanticipated class (one that was not hypothesised in the original model). The SIMCA diagram in Figure 3 illustrates three outlier points which have been excluded from membership to either hypothesised class. In contrast, the left portion (LDA) of Figure 3 shows how three outliers would be treated in a simple linear discriminant analysis. Note that the LDA is "inclusive", *i.e.* it merely defines the "best" plane that divides the two hypothesised classes. Since every point must lie on one side of the separator all points will be "included" with one of the classes. Several distant points belonging to an unanticipated class would be, by virtue of their location relative to the plane, incorrectly assigned to one of the predetermined classes.

Among the insights gained in the SIMCA approach is a knowledge of the variables that best describe each class. This is the result of the separate principal component analysis performed on each hypothesised class. The loadings of the variables on each of the principal components help one to determine which features characterise membership of that class. The number of principal components in addition to the class mean that are required to describe each class help us to "see" the shape of the class in hyperspace. These insights are quite helpful when one is confronted with a classification problem that must be approached in an unsupervised fashion, *i.e.* when one has little knowledge of the correct number of classes to be hypothesised. As the flow diagram Figure 1 shows, the classification analysis is an iterative protocol that permits revision of the model as more insight is gained. In this way "supervised" SIMCA algorithm may be used in pursuing an "unsupervised" learning/exploratory data analysis.

*KNN*

KNN (Varmuza, 1980) is a distance based classification scheme. Each sample can be visualised as a point in some multidimensional space. (Each measured variable defines one axis of this space.) The distances between each point and all others can be computed. This classification technique is also a "supervised" algorithm that requires one to begin with hypothesised classes. A training set consisting of all objects/samples known to belong to various hypothesised classes is constructed. Figure 4 illustrates the process for a simple two-dimensional two category example. The location of a test object (depicted as an asterisk in Figure 4), whose class membership is to be determined, is computed. If one then computes the distance between the test object and all other objects of known class membership, one can prepare a table of the class to which each neighbouring point belongs. The test object may be assigned to the class by a "vote" of the K nearest neighbours. For example, if a test object's three nearest neighbours are all known to belong to "Class A", then the test object is also an "A". If a test object's three nearest neighbours belong to "Class A", "Class A" and "Class C" respectively, then the sample is an "A". While "votes" among any number of nearest neighbours may be considered, theory provides a guide for deciding how large the "electorate" ought to be. One nearest neighbour provides 50% of the membership information, two nearest neighbours provide 75%, three nearest neighbours provide 87.5%, *etc.* Thus, while K (in KNN) may be any number there is a practical means for restricting the size of the electorate. Furthermore, in cases where a tie or ambiguous vote is obtained, one may employ various vote weighting schemes that simplify classification. This scheme is computationally quite simple for a computer. It is non-parametric, requires no complex math functions, and it does not depend upon any prior assumptions of statistical homogeneity or population distribution. The principal weakness of the KNN method is that it does not provide information about the reasons (which variables are the most important) for distinguishing among classes. In this sense it is similar to cluster analysis which identifies separate classes but does not indicate the underlying mechanisms that control their memberships. In spite of these limitations KNN is very useful when it is used in conjunction with other techniques.

*Combining Techniques*

The computer algorithms described here resemble "black boxes". One feeds data into one side of the "box" and retrieves "results" from the other side. Humans are always uncertain about the validity of results that come out of black boxes. This scepticism is healthy and should be encouraged since many of the assumptions implicit in the execution of the algorithms are invisible to the user. It is useful to view these computer algorithms as instruments of reasoning. Adopting this view permits us to formulate strategies for assessing the validity of the output. For example, when one wishes to test performance of modern analytical instruments one performs validity testing on materials of known composition, standard reference materials (SRM's). By analogy one should test algorithm (reasoning instrument) validity with test data. Simulation Test Data sets (STD's) consisting of well characterised data that simulate observations of real experiments are being developed. Such data sets are advantageous since the "truth" is known and the physical model (functional relation) as well as the random error model can be strictly controlled. Carrying the instrument analogy further, one often finds that suitable standard reference materials are not available and one must rely upon validity testing by independent methods. Under these circumstances one assesses accuracy by performing the analysis by several techniques that are based on fundamentally different principles. If the independent techniques provide the answer one gains confidence that the results are accurate. In an analogous fashion one may employ independent classification algorithms to test the accuracy of the classification results. When a consistent picture emerges from all of the various algorithms one may conclude that the results are not an artifact imposed by the selection of a particular algorithm. It should be emphasised that algorithm testing by independent methods must be characterised by the same objectivity required when analytical methods are being evaluated. One may not select from candidate methods the one that provides the expected answer.

50

**Conclusion**

Research in geochemistry and health requires interdisciplinary collaboration. Exploratory data analysis enhances this collaboration and facilitates the cross-disciplinary communication because it permits simultaneous examination of many types of variables. The methodologies of exploratory data analysis are gaining increased acceptance in the scientific community, yet, there remains some controversy about the appropriateness of these methods in science. Critics have called this approach to data analysis "science without hypothesis" and "fishing expeditions". Because of this controversy a brief discussion of the philosophy of this approach is warranted.

Herbert A. Simon (Simon, 1981) has written eloquently about the artificial constructs or models that man has developed to assist him in his understanding of natural phenomena. Simon states,

"The central task of a natural science is to make the wonderful commonplace: to show that complexity, correctly viewed, is only a mask for simplicity; to find pattern hidden in apparent chaos."

If we accept this broad view of the scientist's task, then natural science is a body of knowledge about some class of things, objects or phenomena. The scientist seeks to describe the characteristics and properties that the objects have, and to explain the interactions in the system. In the early stages of observing natural phenomena the scientist assembles qualitative and quantitative observations that may be used to group the objects according to class "behaviours". Assigning objects to classes provides a means of describing complex relationships by minimising small differences so that the larger differences may be summarised and studied. By identifying the "differences that make a difference" the basis for formulating hypotheses is established. Once a picture begins to emerge, the scientist applies intuition and formulates a working hypothesis that suggests additional observations or experiments. The new observations form a basis for accepting, rejecting or revising the hypothesis. This iterative process ultimately leads to a more general theory. Implicit in the hypothesis is the desire to extend the simple taxonomic classification of the objects to a mechanistic description of the underlying principles that permit class distinctions. When the task of science is viewed in this way, it is apparent that the chemometric approach has much to offer. The mathematical tools provided by statisticians are adapted by chemometricians and fashioned into reasoning instruments. When these techniques are properly used they can provide a valuable interpretive aide in summarising existing data, formulating new hypotheses, and designing better experiments.

**References**

Albano, C., Dunn, W., Edlund, U., Johansson, E., Norden, B., Sjostrom, M. and Wold, S. (1978). *Anal. Chim. Acta*, **103**, 429-443.
Anderberg, M.R. (1973). *Cluster Analysis for Applications*. Academic Press, New York.

Cooley, W.W. and Lohnes, P.R. (1971). *Multivariate Data Analysis*. John Wiley and Sons, New York.

Davis, J.C. (1973). *Statistics and Data Analysis in Geology*. John Wiley and Sons, New York.

Erickson, G.A., Jochum, C., Gerlach, R.O. and Kowalski, B.R. (1980). Paper 99, 65th Ann. Mtng. Am. Assoc. Cereal.

Gorsuch, R.L. (1974). *Factor Analysis*. W.B. Saunders Co., Philadelphia.

Harper, A.M., Duewer, D.L., Kowalski, B.R. and Fasching, J.L. (1977). In: B.R. Kowalski (ed.), *Chemometrics: Theory and Applications*, pp.14-52, ACS Symposium Series, No.52, American Chemical Society, Washington, DC.

Kowalski, B.R. (ed.) (1977). *Chemometrics: Theory and Applications*, ACS Symposium Series No.52, American Chemical Society, Washington, DC.

Malinowski, E.R. and Howery, D.G. (1980). *Factor Analysis in Chemistry*. Wiley Interscience, New York.

Massart, D.L., Kaufman, L. and Coomans, D. (1980). *Anal. Chim. Acta*, **122**, 347-355.

Simon, H.A. (1981). *The Science of the Artificial*, MIT Press. Cambridge, MA.

Varmuza, K. (1980). *Anal. Chim. Acta*, **144**, 227-240.

# 6 Dispersion and Deposition of Radionuclides from Chernobyl

H.M. ApSimon, J.J.N. Wilson and K. Simms
*Centre for Environmental Technology*
*Imperial College, London SW7 2PE, England*

## Introduction

When the Chernobyl accident occurred there was a need for rapid assessment of the geographical distribution of contamination to aid decisions on precautionary protective measures. Each country required information not only on its own home territory but throughout Europe to safeguard citizens abroad and regulate food imports. This paper describes some of the problems in assembling and interpreting data in such circumstances, and how computer models have been used to simulate the pathways of radionuclides from the reactor to the site of measurement to check consistency and provide a coherent picture of environmental contamination.

In an unplanned situation such as Chernobyl and with the spanning of many small countries in Europe, the difficulties of collecting consistent data are exacerbated. This was particularly true of the radiological measurements, for which there was no established international system. By contrast the meteorological data required to simulate dispersal of the release is routinely reported with high reliability through the World Meteorological Networks, and was quickly supplied from the UK Meteorological Office. The accident involved releases of a large number of different radionuclides for which nuclide characteristics and behaviour had to be specified. Some of these data, such as decay rates, are well established, but others were completely unknown. Until the Soviet presentation in Vienna in August, 1986, 4 months after the accident, there was very little information on the source itself (and even now there are large uncertainties). Thus the quantity of each nuclide released and its variation with time required educated guesses or deductions from other data. The source terms are of interest in their own right to reactor physicists, and hence are also addressed in this paper.

Thus it is evident that many different data sets are involved in analysis of the Chernobyl accident - the basic radiological measurements, ancillary meteorological data and fundamental reactor and nuclide data. Data which would be available in an organised experiment, such as the amounts released, had to be prescribed by independent means.

**The Radiological Measurement**

Immediately following Chernobyl divergent attitudes emerged between different countries - some reported the highest levels of radioactivity observed, other tried to give representative or average values, and others disclosed no measurements at all. The World Health Organisation collected and disseminated data internationally, and in the UK the National Radiological Protection Board accumulated information which we were grateful to have made available to us. However much of the data on which our early analysis at Imperial College was based was acquired over the telephone, or handwritten on telefax, or by telex, directly from national laboratories whose data was likely to be reliable. Bank holidays and a telecommunication strike in Finland made the task more difficult.

Even then different measurement techniques had to be allowed for. For example the $I^{131}$ released was present partly as iodine vapour and partly attached to atmospheric aerosols; some measurement techniques used filters which only recorded the latter portion which was typically 1/3 to 1/5 of the total. Measurement techniques were rarely defined, but whether the total $I^{131}$ or just the particulate fraction had been observed could usually be determined by considering the ratio of $I^{131}$ to other nuclides such as $Cs^{137}$. The particulate fraction alone would seem anomalously small. Nevertheless many national reports report a mixture of $I^{131}$ measurements indiscriminately! Simultaneous measurements including two different isotopes of the same element such as $Cs^{137}/Cs^{134}$ or $Ru^{103}/Ru^{106}$ also provided a useful check, since they should behave in a similar way in the environment and hence reflect the relative proportions within the reactor at the time of the accident (allowing for decay and any pre-existing background).

Further difficulties arose in the units used to report measurements, compounded by a relatively recent change from Curies, Rads and Rems to Becquerels, Grays and Sieverts (plus all their micro- and milli-subunits). Unfortunately many early measurements had to be discarded because the units were not stated or were ambiguous. Air concentrations were the simplest usually being expressed in Bq m$^{-3}$ of air (but sometimes Bq kg$^{-1}$), but again the sampling period and duration were not always indicated.

Measurements of deposited activity raised great problems. Firstly deposition of material in rainfall gave rise to an extremely patchy distribution of contamination and it was difficult to decide when measurements were taken in hot-spot areas, and when they pertained to low levels in between. Secondly there was no standard way of measuring deposition. Some institutions, such as the Institute for Terrestial Ecology (ITE) in the UK, measured activity deposited on grass; which represents only a fraction (typically ~ 20%) of the total deposition per square metre, and can vary considerably with the type of sward etc. Others measured deposition in soils, but excavated to different depths. Measurements were reported either in units per square metre, or per square kgm (with no indication of kgm m$^{-2}$ in the sample). A few measurements were for artificial surfaces such as tarmac and roads but this was not necessarily clear.

Measurements in foodstuffs introduce additional factors. The wide variety of foods monitored and differing agricultural practices (*e.g.* cows housed indoors or out on pasture) introduced such complexities that at Imperial College we used them only qualitatively to indicate which foods were likely to be most significant and confirm relative levels of contamination over broad geographical regions.

Given these difficulties with the radiological data it is not surprising that there was confusion in obtaining a consistent picture of contamination across Europe.

**Meteorological Measurements and Computer Modelling**

Computer simulations of the dispersal and deposition of radionuclides released according to the meteorological conditions provided potential tools to unravel the radiological data. These however require large amounts of meteorological data. The MESOS model (ApSimon *et al.*, 1985) developed at Imperial College to study a large number of hypothetical nuclear accident scenarios for risk assessment studies, fortunately uses only data routinely reported at synoptic stations. This data is used in weather forecasting models and hence was readily available. Thus when the Nuclear Installations Inspectorate (NII) asked us to apply MESOS, the UK Meteorological Office very quickly and efficiently supplied the synoptic data in exactly the same format as for our previous hypothetical studies. Handling this data, albeit for about 800 irregular spaced stations across Europe, was therefore relatively straightforward.

The essential data included surface pressures, cloud cover, temperature and observations of present weather indicating the nature of any precipitation (*e.g.* snow, continuous drizzle, heavy continuous rain) at three hourly intervals. Data handling programs already developed to take into account the topographical characteristics of land, sea and mountain barriers in a simple manner, were quickly applied to pre-process the data for a regular grid of latitude and longitude cells over a map area from 10°W to 35°E and 62°N to 36°N. Some *radiosonde* data giving vertical profiles of wind and temperature within the USSR were also very useful, showing a high level inversion at about 300 metres when the accident occurred and night time surface inversions.

However some adaptions were required to the main MESOS programs simulating the progressive spreading of material, since they had not been designed for use in a real situation. Thus for example they had hitherto estimated only the total integrated contamination, and not a time sequence of its occurrence.

**The MESOS Model**

The model is a Lagrangian trajectory model. It follows the passage and dilution of a sequence of puffs released at 3-hourly intervals throughout the release period, taking into account meteorological conditions en route. Trajectories are based on 1,000 mb geostrophic windfields deduced from gridded surface pressure data, with some adjustment for the vertical spread of material over the boundary layer of the

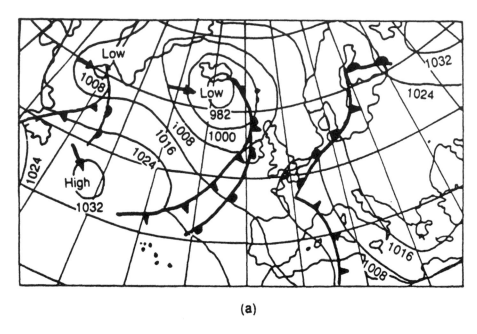

(a)

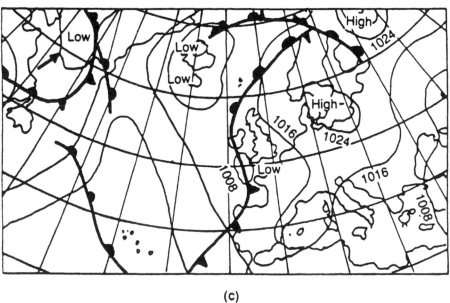

(c)

**Figure 1.** *Midday Weather Charts 26.4.86 to 5.5.86.: (a) Saturday 26.4.86.;*
*(b) Wednesday 30.4.86.; (c) Friday 2.5.86.; (d) Monday 5.5.86.*

56

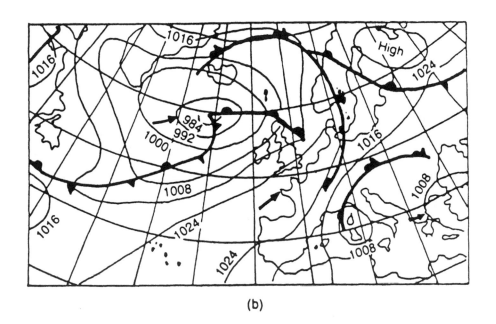

(b)

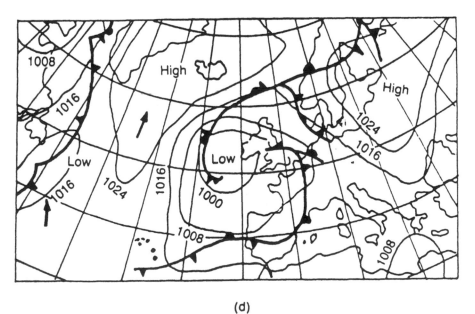

(d)

57

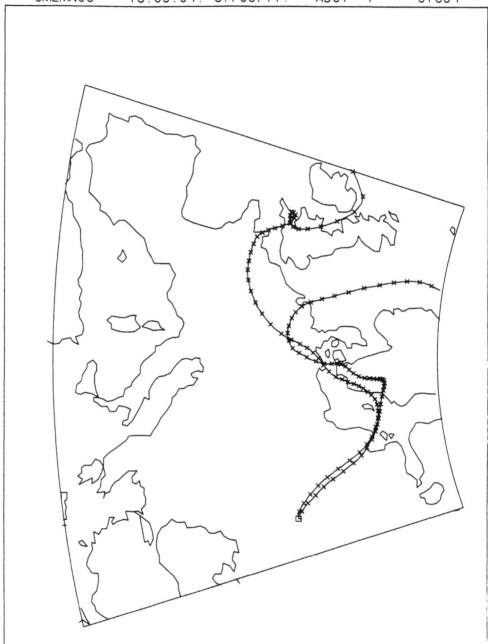

PLOT COMPLETE.   PENS USED - ⧓

**Figure 2**. *Trajectories of material released at 9.00 hours and 12.00 hours on 26th April 1986.*

atmosphere. Deposition of material on the ground depends on the character of the radionuclides involved. It is allowed for in the model through dry deposition from contaminated air at the surface, and may be much enhanced when scavenging occurs in precipitation; a simple wash-out model is used for the latter with a wash-out coefficient depending on rainfall rate. This is deduced from the observation of "present weather" associating effective rainfall rates with each weather type. For each 3-hour period of the release the material is assumed to spread out horizontally between and along the trajectories of the discrete puffs whose history is followed in detail.

## Simulation of the Dispersal of the Chernobyl Release

The USSR had disclosed that the accident had occurred at 1.23 hours local time on 26th April (equivalent to 21.33 hours on 25th April GMT). From observations in Scandinavia it was apparent that a wide range of radionuclides was being released and that a serious accident had occurred.

The first step was to estimate the trajectories of air parcels originating from Chernobyl at 3-hourly intervals. The meteorological situation is shown in Figure 1. Initially material travelled towards Scandinavia. Conditions then became very stagnant for a couple of days over the Ukraine and North Eastern Europe with a front giving some rain over Scandinavia. By Wednesday, 30th April a tongue of clean air from the Atlantic was moving across France and England towards Denmark. This change had a marked effect on subsequent dispersal of the cloud, giving rise to a high pressure region that moved eastwards across the north of Europe. Contaminated air stagnated and travelled clockwise round this high pressure area, reaching Britain on 2nd May, seven days after the start of the release. Figure 2 shows calculated trajectories illustrating this. Some material then circulated round the low pressure region to the west of Britain, while material from Central Europe was funnelled up northwards over the North Sea and Norway. Trajectories originating on 26th and 27th April are highly variable, and only material within a 3-6 hour segment of the release travelled over the UK. Any release from noon on 27th to 18.00 hours on 28th April appears to have travelled eastwards across the USSR. Subsequently trajectories indicate material released travelled southwards towards Greece and then up over Scandinavia again in the early days of May.

In former applications of the MESOS model trajectories had been followed for 96 hours at the most. In the case of Chernobyl this was considerably extended to 8 days or more. Nevertheless the trajectories still appeared consistent with observations, when modelling results were checked against measurements collected from various sources across Europe.

## Characteristics of the Release

Information on the reactor (an RBMK Soviet design with graphite moderator and water as a coolant) was also required. Pre-calculated inventories were available (*e.g.* US Nuclear Reactor Regulatory Commission Reactor Safety Study, 1975) which

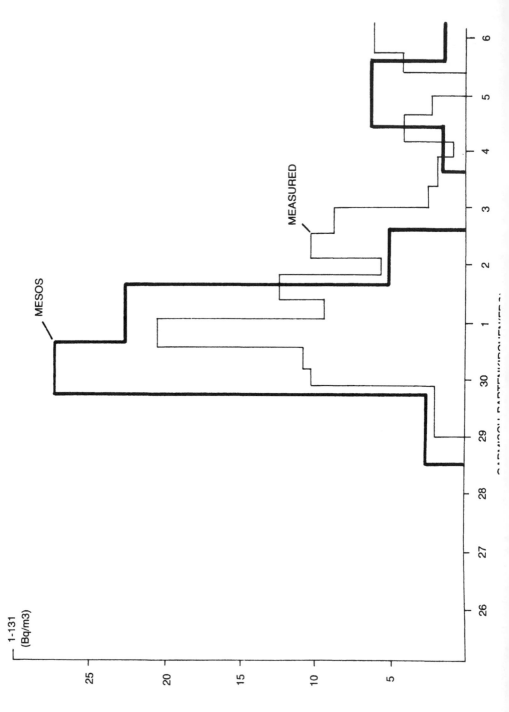

**Figure 3.** *Example of comparison between calculated values and measurements of* $I^{131}$ *in air.*

gave a good indication of the quantities of important nuclides inside the reactor at the time of the accident. It was assumed that the noble gases (*e.g.* $Xe^{133}$) would escape relatively easily in the early hours of the release. The volatile nuclides including $I^{131}$ and $Cs^{137}$ would also be released in relatively large amounts. Other nuclides such as $Sr^{90}$ would be retained more within the fuel matrix and would be transported with particles of fuel which were more likely to be deposited closer to the reactor by gravitational settling. (Some "hot particles" did travel as far as Scandinavia, possibly having been carried aloft in convective storms.)

The initial explosions carried material up to 1,000 metres or more, and subsequent heat in the release gave rise to an elevated plume. Since there was an elevated inversion at 3,000 metres, vertical dispersion allowed considerable dilution initially.

To start with dispersal of nominal amounts of $I^{131}$ and $Cs^{137}$ was simulated separately for each day to investigate release scenarios. The model was used to calculate daily average air concentrations and dry and wet deposition over a European network of grid cells. The air concentrations were compared with measurements (see for example Figure 3) and quantities released normalised to obtain reasonable agreement. In this way it was possible to make an approximate estimate of quantities released. Based on a release scenario postulated at the time, with an initial peak in the release sustained over the first two days thereafter reducing substantially over the next two days, observed levels of contamination over northern and western Europe appeared consistent with a release of 15 to 25% of the $I^{131}$ and $Cs^{137}$ in the core (that is up to 7.5 x $10^{17}$ Bq of $I^{131}$ and 5.0 - 6.0 x $10^{16}$ Bq of $Cs^{137}$). These results, given in a presentation at the NII on 20th May, are in good agreement with values eventually reported by the USSR in Vienna at the end of August. However this agreement may be partly fortuitous; since the USSR estimates were based on deposited $\gamma$ just within the USSR which would include fuel debris settling out close to the source, whereas our estimates are based on deposition travelling further afield beyond the USSR. In any case we presented our estimates with caution and caveats not only because of modelling uncertainties but also due to:

(i) the uncertainties already described in measurements and measuring techniques used (particularly for $I^{131}$), and incomplete information about the time periods to which they referred;

(ii) lack of information over the USSR to indicate the amounts released on 27th and 28th April (the USSR have since reported that relatively little was emitted during this time).

It was also apparent that levels of contamination calculated for Greece and south eastern Europe were too low. Trajectory analysis and times of arrival of material in Greece implied that the release was far more prolonged into the early days of May. This was later confirmed by the USSR disclosure in Vienna that there had been a second peak in the release from 2nd to 6th May when material was heaped over the damaged reactor core.

Figure 4 shows the simulated spreading of contamination across Europe from our early calculations; the levels represent the accumulated deposition of $Cs^{137}$ by

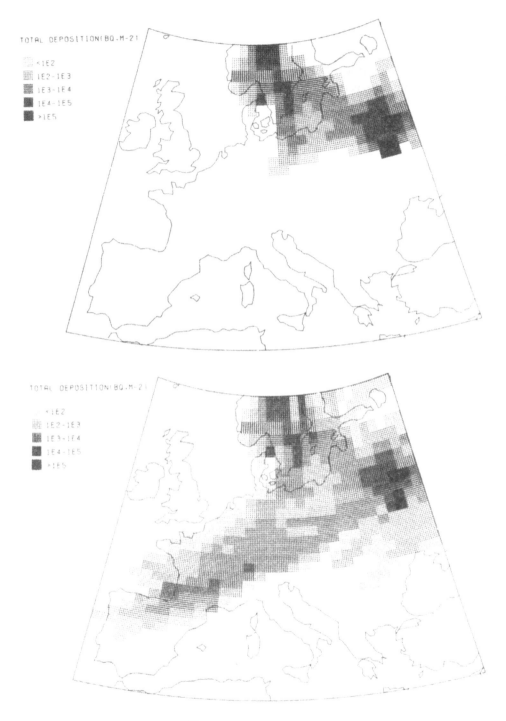

**Figure 4**. *Cumulative Total $^{137}$Cs Deposition: (a) 21.00 25.4. - 09.00 29.4.;*
*(b) 21.00 25.4. - 09.00 1.5.; (c) 21.00 25.4. - 09.00 3.5.; (d) 21.00 25.4. - 09.00 8.5.*

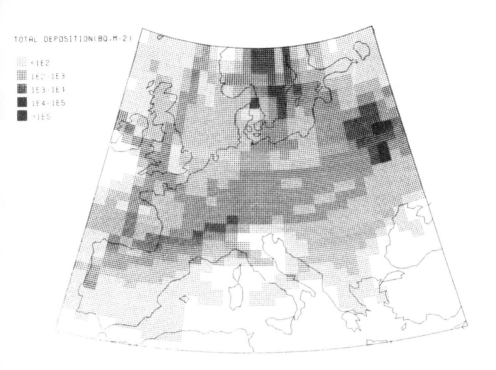

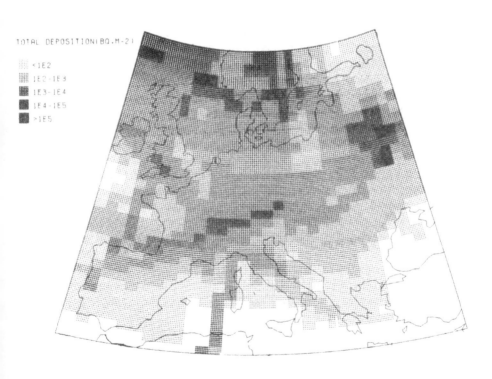

different dates, this being the most significant factor for long-term dose commitments to the population.

These maps were used in combination with the population distribution to estimate the collective dose commitment over the next 50 years to the European population outside the USSR - bearing in mind uncertainties, a tendency towards pessimism in dose factors used in risk analysis and any reduction which may have resulted from the introduction of control measures.

For the 550 million people in Europe outside the USSR the calculated result was x 2.0 $10^5$ man Sv from Cs isotopes - allowing for approximately equal dose commitments from external irradiation and ingestion and including an extra contribution for $Cs^{134}$. This amounts to an average dose commitment per individual of 0.4 mSv. In severe hot spot areas outside the USSR with up to 50 times the average estimated deposition of ~ 2 kBq $m^{-2}$ of $Cs^{137}$ and $Cs^{134}$, the average individual dose commitment estimated on the same basis is 10 to 15 mSv.[*] By comparison the maximum permitted dose limit over a lifetime for a member of the public is 70 mSv. The additional cancers will be impossible to detect from epidemiological studies against the normal occurrence.

All this information was made available as requested to such bodies as the World Health Organisation and the European Commission.


## Detailed Studies of Deposition Over England and Wales

The picture presented above gives a very broad picture of contamination and does not indicate localised hot spots of deposition. Attention was therefore given to obtaining a more detailed picture of deposition of $Cs^{137}$ over England and Wales. This required more precise data, particularly on precipitation.

Observations at CEGB power stations indicated that the material passed across Britain in a steep ripple, rising sharply and falling again at each location in a matter of a few hours. Although it was a complex frontal situation with contaminated air feeding into a steep vertical ascent and yielding precipitation, a simple analysis of where and when this rain occurred as observed by the weather radar network of the Meteorological Office, gives a good indication of deposition of Cs. Each weather radar produces data out to a maximum of 200 km; a composite map of rainfall intensity is produced from the network every 15 minutes with a 5 x 5 km grid resolution. This is a substantial amount of data. Again a model (RAINPATCH) had been developed at Imperial College to use weather radar data in analysis of hypothetical nuclear accidents, and it was brought into use with the data readily made available to us by the Meteorological Office.

The long-range trajectory analysis of the overall travel across Europe to the UK could not be relied upon to pinpoint with sufficient accuracy the timing of passage of

---

[*] Following the information made available in Vienna at the end of August revised calculations were performed. These imply higher levels over south-eastern Europe towards Greece, and slightly lower levels over western Europe, yielding 1.6 x $10^5$ man Sv from Cs isotopes to the European population outside the USSR, an average of ~ 0.35 mSv per individual.

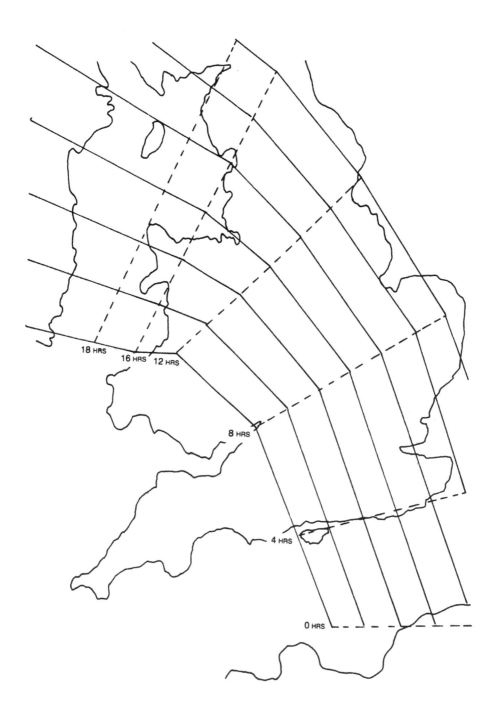

**Figure 5**. *Time of passage of Chernobyl cloud.*

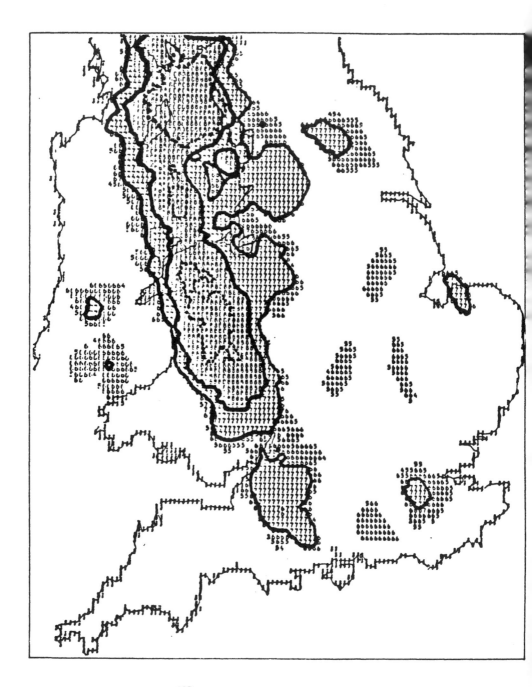

**Figure 6**. *Deposition of Cs^137 over England and Wales as calculated with RAINPATCH model.*

material. Trajectories were therefore re-calibrated to the times of arrival at CEGB power stations. The airborne radioactivity was then treated as overlapping puffs advancing in a line across the UK (see Figure 5).

RAINPATCH was then used to simulate differential wash-out and deposition of material from the cloud where it intercepted rain as indicated by the radar data. The resulting estimated map contamination is shown in Figure 6, the two solid contours indicating levels differing by an order of magnitude, and the innermost contour higher by a further factor of 3. It is evident the levels were markedly higher over North Wales, Cumbria and the Isle of Man.

These results are in good agreement with measurements made by the Institute of Terrestial Ecology (ITE) even to small areas of lower contamination in East Anglia. Unfortunately they took several months to produce, although potentially they could have been available almost as the rain occurred.

## Conclusions

Our experience in using a meteorological model to interpret information following the Chernobyl accident, has not only been an interesting and we hope useful exercise, but has also indicated requirements for future accident situations. The first is the necessity for much improved communications and information, both about the release and its variation in time and radiological measurements. There is a need to establish common standards and procedures for these radiological measurements, and also for their rapid reporting in standard units and formats. The same applies to establishing common derived emergency reference levels and intervention levels for control measures as far as possible. Because of the patchy nature of precipitation and its potential importance for deposition of radionuclides, good data on when and where rain or snow occurs is valuable. This has been shown to be feasible using weather radar for which the network is well advanced over England and Wales with plans for its extension to Scotland.

We feel that we have successfully demonstrated that atmospheric dispersion models can be a useful tool for analysis of nuclear accident consequences. There is now much interest in using such techniques in conjunction with forecasting analyses to provide supporting services and information rapidly in the event of any future accident of this sort, both on a national and international basis.

## References

ApSimon, H.M. *et al.* (1985). Long range atmospheric dispersion of radio-isotopes. Part I: the MESOS model; Part II: Application of the MESOS model. *Atmospheric Environment*, **19**, 99-125.

US Nuclear Regulating Commission Reactor Safety Study (1975). WASH 1400, NUREG 751014. Washington, DC.

# 7 Relationships between the Trace Element Status of Soils, Pasture and Animals in relation to Growth Rate in Lambs

N.F. Suttle
*Moredun Research Institute, Edinburgh EH17 7JH, Scotland*

**Summary**

*Nine "animal factors" influencing the relationship between trace element status in soils, pastures and grazing animals create difficulties in predicting growth-limiting deficiencies in lambs from soil or herbage composition. Examples are presented of the influence of four factors: of herbage Mo and animal genotype on Cu absorption; of initial reserves on later Cu, vitamin B12 and Se status and of ingested soil Co on B12 status.*

**Introduction**

Trace element deficiencies have been known to adversely affect the health and productivity of grazing livestock for so long now that the uninitiated might expect that such problems could be readily predicted and the relative merits of soil, plant and animal indices as the basis for prediction would be clear. As far as soil indices are concerned, claims have been made for geochemical mapping (Plant and Stevenson, 1985) and soil survey approaches (Ure, 1985). In both the UK (Thornton, 1983) and the USA (Kubota, 1976), Mo-induced Cu deficiency in cattle has been clearly associated with Mo-rich outcrops. For plant indices, geographical distribution of Co-deficiency in grazing animals in the USA has been indicated by maps showing Co concentrations in legumes (Kubota, 1968). Such general associations might be useful to governments for assessing research priorities and state agriculturalists for determining development priorities for funding. At the end of the day, however, such predictions have to be applied to the individual farm and the farmer wishes to know whether the purchase of trace elements will produce a worthwhile financial return. Can knowledge of the trace element status of soils, pasture or animals best assist him or must all be incorporated in the predictive model? If all are required, then prediction itself will become a costly exercise.

69

## Principal Components

Many factors will determine whether or not a soil of particular trace element composition will give rise to deficiency (or toxicity) problems in grazing animals. Figure 1 outlines those factors that relate to the plant/animal and soil/animal interfaces. I have treated the important soil/plant interface as a "black box" because my knowledge of it extends only to an awareness that estimates of availability of soil-borne elements to the plant is as crucial as their availability in plants to animals. At the plant/animal interface appetite, selective grazing, availability of ingested elements during digestion and the capacity of the animal to absorb and store what becomes available will all influence the relationship between herbage and animal status; they must shape our efforts to devise predictive models of animal responses. The paper will illustrate the importance of two of these factors (availability and capacity to absorb) together with initial reserves and a further important interface (soil/animal) and how all these parameters might be accommodated in predictive models.

One example will suffice to show the limitations of prediction from trace element analysis of soils alone. Maps of soil Cu concentrations in Scotland indicate that the SW and Western Isles might spawn problems of Cu deficiency (SAC/SARI, 1982).

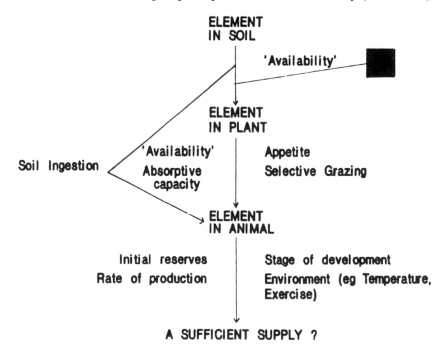

**Figure 1.** *Many factors influence the flow of trace elements from the soil via the pasture to the grazing animal and whether or not a particular supply to the tissues will result in loss of production. Factors influencing the uptake of elements by the plant are (black box) not considered in detail.*

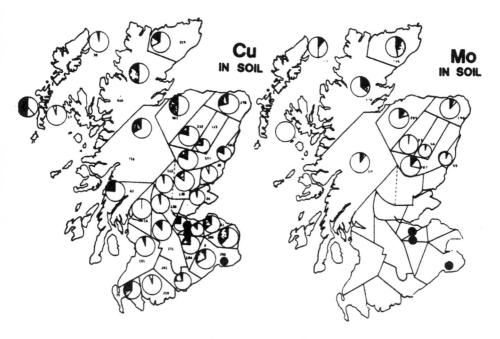

**Figure 2**. *Assessments of risk of Cu deficiency in Scotland from extractable Cu or Mo in soil. Adequate (white), borderline (stippled) or deficient (black) status is indicated by Cu >1.5, 0.8-1.5 and <0.8 mg/kg and by Mo <0.05, >0.05-0.1 and >0.1 mg/kg soil respectively (from SAC/SARI, 1982). Three incidents of hypocuprosis are shown (•) (Whitelaw et al., 1979; 1982; Woolliams et al., 1986b).*

The same authority, knowing that Mo was an important determinant of the availability of Cu in pastures to animals and that drainage conditions determined Mo availability to pastures, produced complementary maps of extractable soil Mo and soil drainage which showed that problems were likely to be most prevalent in the NE of Scotland (Figure 2). Mo-induced Cu deficiency in growing lambs on improved hill pastures has been reported at three sites, all outside the areas of highest risk predicted from information on soil Cu or Mo. The incidents occurred on experimental farms (Whitelaw *et al.*, 1981; 1982; Woolliams *et al.*, 1986 a,b), but surveys of herbage Mo on other improved hill pasture sites in Scotland suggested that problems might also be widespread (C.C. Evans, HFRO - personal communication) outwith the predicted areas in Figure 2.

The Cu-responsive conditions arising from hill pasture improvement (Figure 2) were probably attributable to the combination of a change in soil pH after liming, an improvement in pasture management and the presence of a susceptible breed of sheep: trace element composition of the soil was not a major factor. No uptake by the pasture increases with soil pH (Burridge *et al.*, 1983) and the greener, heavily

**Table 1.** *Contrasting incidence of hypocuprosis in Scottish Blackface (B) and Welsh Mountain (W) lambs in the first two years after grazing the same improved hill pastures. The signs of hypocuprosis were mortality (largely from infections and swayback, expressed as a proportion of lambs born), anaemia (given as a proportion of surviving lambs with haemoglobin <80g/l sometime between 6-24 weeks of age) and growth retardation (difference in growth rate (g/d) between untreated and Cu-treated lambs between 6-24 weeks): from Woolliams et al., 1986 a,b).*

|                      | Year 1 |      | Year 2 |      |
|----------------------|--------|------|--------|------|
|                      | B      | W    | B      | W    |
| Mortality            | 0.45   | 0.04 | 0.28   | 0.06 |
| Anaemia              | 0.04[*]| 0.00 | 0.24   | 0.00 |
| Growth retardation   | 9.2[*] | 8.5  | 40.9   | 7.6  |

[*] Removal of the most vulnerable lambs by death may underestimate the severity of hypocuprosis as indicated by anaemia and growth retardation in B, particularly in Year 1.

stocked pasture is higher in S (Whitelaw *et al.*, 1981). Together these two elements can lower Cu *availability* by a factor of 3 following modest increases in concentrations of 3 mg Mo and 2 g S/kg herbage DM (Suttle, 1986). The changes at one site were only sufficient to induce hypocuprosis in a breed with a particularly low efficiency of absorption, the Scottish Blackface (Table 1); the Welsh Mountain was largely unaffected by the management change (Woolliams *et al.*, 1986 a,b), indicating the importance of *capacity to absorb* (Figure 1) as a factor influencing the development of Cu- responsive disorders. If any relationship exists between trace element composition of Scottish soils and the incidence of growth retardation, the relationships clearly cannot be extrapolated to the Welsh uplands where the resistant breed predominates although they may be relevant to areas populated by the North County Cheviot (*cf.* Whitelaw *et al.*, 1983).

A recent survey by Leech and Thornton (1987) has suggested that S deposition from industrial pollution may be an important determinant of incidence of Cu deficiency in cattle in the UK, adding a further obstacle to prediction of disorder from soil analysis alone.

## Quest for a Database

Recently observed links between trace element deficiencies and resistance to infection indicate that the search for predictive models must go on. Growth retardation at two improved hill sites was preceded by heavy losses due to miscellaneous infections (Whitelaw *et al.*, 1981; Woolliams *et al.*, 1986a). At one site the losses in Cu deficient lambs due to infection was repeated in a third (successive) year (Suttle *et al.*, 1987a). Co and Se deficiencies may also influence susceptibility to infection (Paterson and MacPherson, 1987; Turner *et al.*, 1985), and in economic terms the loss of lambs from infections is more costly than a marginal reduction in growth rate.

Growth retardation and susceptibility to infection are equally sensitive to Cu deficiency (Suttle *et al.*, 1987b) but the former is easier to monitor. Coordinated growth response trials were therefore set up to see how frequently performance was retarded by Cu deficiency in Scottish hill lambs on improved pastures and whether losses could be predicted from measures of Cu status in the soil, herbage and grazing animal (Suttle *et al.*, 1986). Since the same management change was believed to increase the incidence of Co- and Se-responsive growth retardation in lambs (SAC/SARI 1982), Co and Se status of the soil, plant and animal was also monitored and growth responses to supplementation with Co and Se were sought, in the second and third years respectively, of the three-year study (Suttle *et al.*, 1986). The Scottish trials included three groups, *i.e.* a totally unsupplemented control group (C), a group treated with all three elements of interest (T3) and a third given all but one (T2), the missing element being that to which growth responses were sought in a given year. The comparison T3-T2 gave the response to one element while T2 *versus* C in any year gave an indication of the combined effect of the two elements not specifically under test that year. Thus in the first (Cu) year the combined effect of Co and Se was also indicated.

The results in year 1 indicated that deficiencies of the elements had not caused widespread growth retardation. Only 1/21 farms gave a significant response in liveweight gain to treatment (1.3 g CuO needles at six weeks), the average improvement being a meagre $4 \pm 1.8$ g/d. The combined Co and Se treatment also failed to improve growth. The lack of response to Co was at variance with predictions made from analysis of the extractable Co in Scottish soils, (SAC/SARI, 1982) and from Cu, Mo, Co and Se analysis of herbage, all of which indicated a high risk of animal requirements not being met on some farms (SAC/SARI, unpublished data).

Blood analyses indicated why responses were not obtained. The plasma Cu concentrations were subnormal (<9 μmol/L) on only 3/21 farms in May and they showed a spontaneous recovery through to August. Plasma B12 (a marker for Co status) and glutathione peroxidase (GSHPx) (for Se status) showed opposite trends to plasma Cu, starting high and declining through to August (Suttle *et al*, 1986), though never to concentrations currently considered to indicate risk of growth retardation (250 *pg*/mL and 30 U/g Hb respectively; Suttle (1986).

**The Initial Reserve**

The lack of a response to Cu between May and August did not rule out the possibility of an earlier constraint on growth. We were surprised to find a correlation between

73

**Table 2.** *Relationships (d.f. 16-19) between initial trace element status and subsequent growth or trace element status in grazing lambs, illustrating the possible importance of initial reserves in determining risk of deficiency (Suttle et al., 1987).*

Liveweight gain (g/d)  = 156 - 0.0041 ESOD in May  r = 0.68
(May-August)
Glutathione peroxidase = 36.0 + 0.56 GSHPx in May  r = 0.88
(GSHPx, U/g Hb, August)
Plasma B12      = 250 + 0.509 B12 in May   r = 0.55
(*pg*/mL, August)

activity of a Cu enzyme in the erythrocyte, superoxide dismutase, on the day of treatment and subsequent growth rate that accounted for 36% of the variation in LWG between farms (Table 2). It seemed possible that *perinatal Cu supply* had set early limits on potential growth (Suttle *et al.*, 1986), giving an example of the influence of initial reserves (Figure 1) on loss of production.

The data for Co and Se also indicated a marked effect of the perinatal supplies of these two elements. The principal determinant of plasma B12 and blood GSHPx in August was in each case the respective value in May (Table 2). Herbage analysis revealed little or no influence of Co and Se supply between May and August and their measurement was of no predictive value. Extractable soil Co was correlated with plasma B12, however, (Suttle *et al.* 1987) and it is possible that it gives a measure of Co supply throughout the year and particularly during the important perinatal period. Other studies showed that GSHPx activities in spring were significantly higher in single than twin lambs (Suttle, Jones, Woolliams and Woolliams, unpublished data), indicating perhaps the competition between lambs for the all-important early *maternal supply of Se*.

Under these circumstances, any managerial step during winter which influences trace element supply to the pregnant ewe would be likely to influence the trace element responsiveness of the lamb and must be included in the predictive model (Figure 1).

### The Soil/Animal Interface

The inadvertent *ingestion of soil* as a pasture contaminant may influence the trace element status of the grazing animal in either positive or negative ways and must therefore be considered in models for the prediction of risk of trace element deficiency (Figure 1).

Soils contain far more Co than the pastures they support and sufficient to influence the synthesis of B12 by rumen microbes *in vitro* (McDonald and Suttle,

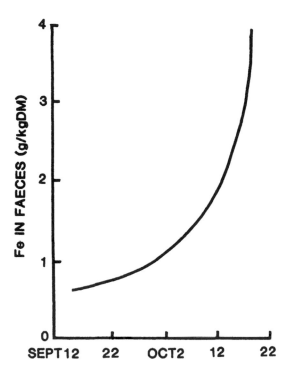

**Figure 3**. *Faecal Fe concentrations (y, g/kg DM) are higher in autumn than in summer and can increase exponentially with time in sheep grazing permanent pasture. (Relationship of y and time (x, days) is a plot of the reciprocal of 1.79-0.04x using data for four years from 1978-1981.*

1986) and *in vivo* (Brebner *et al.*, 1987) at doses likely to be encountered as a pasture contaminant by most grazing animals. Iron concentrations in faeces of grazing ewes at Moredun Research Institute vary substantially during the year and can increase exponentially as progressively more soil is ingested (Figure 3). The maximum Fe values attained in October correspond to those found after adding 3.5% of the Moredun soil to a highly digestible semi-purified diet (Brebner, 1986), but rumen B12 synthesis was increased by adding as little as 1.75% soil to such diets. The response in ruminal B12 synthesis was correlated with extractable soil Co intake (Figure 4; Brebner *et al.*, 1987). Soil ingestion may therefore help to explain the unexpectedly low incidence of Co-responsive growth retardation in the field trials. Ingestion of soil over winter may have contributed to the high initial B12 status of the lambs in the spring, having a decisive effect in preventing Co from becoming limiting for growth by the autumn.

The negative effect of soil ingestion is on Cu status. Soil ingestion at 10% of dry matter intake reduces Cu availability drastically under some experimental conditions (Suttle *et al.*, 1975; 1984) though not all (Langlands *et al.*, 1984). The influence of soil ingestion on Cu-responsive conditions is possibly shown by the scarcity of cases

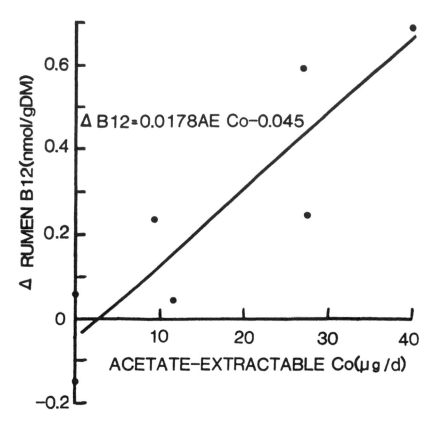

**Figure 4.** *Relationship between acetate extractable Co intake and change in vitamin B12 concentration in rumen liquor in sheep given a low Co diet supplemented with various soils for 21 days (from Brebner et al., 1987).*

of swayback in years of high snow cover when soil ingestion is abruptly reduced. The low initial Cu status of the Scottish lamb in spring may reflect the high ingestion of soil Fe by the pregnant ewe over winter. The inhibitory effect of ingested soil on Cu absorption has been attributed to the trapping of sulphide as FeS in the rumen (Suttle *et al.*, 1984). For this to happen, the soil Fe must enter solution and the effect of soil ingestion is unlikely to be predicted from total Fe analysis because most of the Fe is present in insoluble forms. Sulphide trapping capacity is related to the extractability of soil Fe in pyrophosphate (Brebner, Thornton and Suttle, unpublished data), and the measurement of extractable Fe may enable the inhibitory effects of soil ingestion on Cu availability to be predicted.

It is clearly insufficient to have only qualitative information on the influence of soil ingestion. Effects must be quantified if at all possible and the measurement of extractable Co and Fe offers some hope that they may contribute to the precision of models for predicting risk of Co and Cu deficiency under grazing conditions.

## Conclusions

The complete model for the factors which determine trace element status of grazing sheep (and cattle) is so complex (Figure 1 oversimplifies the situation) that it seems unlikely that trace- element responsive conditions will be predictable on a farm-to-farm basis from analyses of single or multiple constituents in soil. Availability of the ingested elements, whether as agonists or antagonists, in and on the herbage and the initial trace element status of the animal are of major importance, perturbing the relationship between trace element status in the soil and the grazing animal. Herbage trace element analysis may be an unreliable index when status in the animal is influenced greatly by events taking place when herbage is not the major source of nutrients (winter). If geochemistry has a role, it is more likely to be found in the prediction of general risk over large areas: even here, indices of availability, either of essential element (extractable Co) or antagonist (extractable Mo), should be employed. With the latter, the ratios of agonist to antagonist (*e.g.* Cu:Mo) may be of greater predictive merit than their separate values (*cf.* Figure 2). If there are substantial genetic differences between breeds of species in susceptibility to trace element deficiency, extrapolation of soil- or herbage-based predictions from one species to another obviously will be imprecise.

## References

Brebner, J. (1986). The role of soil ingestion in the trace element nutrition of grazing livestock. Ph.D. Thesis, University of London.

Brebner, J., Suttle, N.F. and Thornton, I. (1987). Assessing the availability of ingested soil cobalt for the synthesis of vitamin B12 in the ovine rumen. *Proc. Nutr. Soc.*, **46**, 66A.

Burridge, J.C., Reith, J.W.S. and Berrow, M.L. (1983). Soil factors and treatment factors affecting trace elements in crops and herbage. *Br. Soc. Anim. Prod. Oc. Publ. No.7*, pp.77-86.

Kubota, J. (1968). Distribution of cobalt deficiency in grazing animals in relation to soils and forage plants of the United States. *Soil Sci.*, **106**, 122-130.

Kubota, J. (1976). Molybdenum status of United States soils and plants. In: W.R. Chappell and K.K. Petersen (eds.), *Molybdenum in the Environment*, Chapter 6, pp.555-581. Marcel Dekker Inc., New York.

Langlands, J.P., Bowles, J.E., Donald, G.E. and Smith, A.J. (1982). The nutrition of ruminants grazing native and improved pastures V. Effects of stocking rate and soil ingestion on the copper and selenium status of grazing sheep. *Aust. J. Agric. Res.*, **33**, 313-320.

Leech, A.F. and Thornton, I. (1987). Trace elements in soils and pasture herbage on farms with bovine hypocupraemia. *J. Agric. Sci., Camb.*, **108**, 591-597.

McDonald, P. and Suttle, N.F. (1986). Abnormal fermentations in continuous cultures of rumen micro-organisms given cobalt- deficient hay or barley as the food substrate. *Br. J. Nutr.*, **56**, 369-378.

Paterson, J.E. and MacPherson, A. (1987). Microbicidal activity of neutrophils of cobalt-deficient and repleted calves. *Proc. Nutr. Soc.*, **46**, 167A.

Plant, J. and Stevenson, A.G. (1985). Regional geochemistry and its role in epidemiological studies. In: C.F. Mills, I. Bremner and J.K. Chesters (eds.), *Trace Element Metabolism in Man and Animals*, pp.900-906. Commonwealth Agricultural Bureaux, Slough, UK.

SAC/SARI (1982). Trace Element Deficiency in Ruminants. Report of a study group of the Scottish Agricultural Colleges and Research Institutes, Edinburgh.

Suttle, N.F. (1986a). Copper deficiency in ruminants; recent developments. *Vet. Rec.*, **119**, 519-522.

Suttle, N.F. (1986b). Problems in the diagnosis and anticipation of trace element deficiencies in grazing livestock. *Vet. Rec.*, **119**, 148-152.

Suttle, N.F., Abrahams, P. and Thornton, I. (1984). The role of a soil x dietary sulphur interaction in the impairment of copper absorption by ingested soil in sheep. *J. Agric. Sci., Camb.*, **103**, 81-86.

Suttle, N.F., Jones, D.G., Woolliams, J.A. and Woolliams, C. (1984). Growth responses to copper and selenium in lambs of different breeds on improved hill pastures. *Proc. Nutr. Soc.*, **43**, 103A.

Suttle, N.F., Jones, D.G., Woolliams, J.A. and Woolliams, C. (1987). Copper supplementation during pregnancy can reduce perinatal mortality and improve early growth in lambs. *Proc. Nutr. Soc.*, **45**, 18A.

Suttle, N.F., Jones, D.G., Woolliams, J.A. and Woolliams, C. (1987). Prediction of disorder in Cu-deficient lambs from different genotypes. In: C. Keen and B. Lonnerdhal (eds.),*Trace Element Metabolism in Man and Animals*. Plenum Press (in press).

Suttle, N.F., Thornton, I. and Alloway, B.J. (1975). An effect of soil ingestion on the utilisation of dietary copper by sheep. *J. Agric. Sci. Camb.*, **83**, 249-254.

Suttle, N.F., Wright, C., MacPherson, A., Harkess, R., Halliday, G., Miller, K., Phillips, P. and Evans, C. (1986). How important are trace element deficiencies in lambs in improved hill pastures in Scotland? *Proc. 6th Int. Conf. Prod. Dis. in Farm Animals*, pp.100-103. Veterinary Research Laboratory, Stormont, NI.

Thornton, I. (1983). Soil-plant-animal interactions in relation to incidence of trace element disorders in grazing livestock. *Br. Soc. Anim. Prod. Oc. Publ.*, No.7, pp.39-50.

Turner, R.J., Wheatley, L.E. and Beck, N.F.G. (1985). Stimulatory effects of selenium on mitogen responses in lambs. *Vet. Immunol. Immunopathol.*, **8**, 119-124.

Ure, A.M. (1985). Soil data in the assessment of geochemical and environmental influences on trace element supply. In: C.F. Mills, I. Bremner and J.K. Chesters (eds.) *Trace Element Metabolism in Man and Animals*, pp.906-909. Commonwealth Agricultural Bureaux, Slough, UK.

Whitelaw, A., Armstrong, R.H., Evans, C.C. and Fawcett, A.R. (1979). A study of the effect of copper deficiency in Scottish Blackface lambs on improved hill pasture. *Vet. Rec.*, **104**, 455- 460.

Whitelaw, A., Fawcett, A.R. and MacDonald, A.J. (1982). Cupric oxide needles in the prevention of swayback. *Vet. Rec.*, **110**, 522.

Whitelaw, A., Russel, A.J.F., Armstrong, R.H., Evans, C.C. and Fawcett, A.R. (1983). Studies in the prophylaxis of induced copper deficiency in sheep grazing reseeded hill pastures. *Anim. Prod.*, **37**, 441-448.

Woolliams, C., Suttle, N.F., Woolliams, J.A., Jones, D.G. and Wiener, G. (1986a). Studies on lambs from lines genetically selected for low and high copper status. 1. Differences in mortality. *Anim. Prod.*, **43**, 293-301.

Woolliams, J.A., Woolliams, C., Suttle, N.F., Jones, D.G. and Wiener, G. (1986b). Studies on lambs from lines genetically selected for low and high copper status. 2. Incidence of hypocuprosis on improved hill pasture. *Anim. Prod.*, **43**, 303-317.

# 8 A Study of Environmental Geochemistry and Health in North East Scotland

P.J. Aggett[*],
*Department of Child Health, University of Aberdeen, Scotland*
C.F. Mills,
*The Rowett Research Institute, Bucksburn, Aberdeen,Scotland*
A. Morrison and M. Callan,
*NERC Unit for Thematic Information Systems, Swindon, Wiltshire, England*
J.A. Plant, P.R. Simpson and A. Stevenson,
*British Geological Survey, Grays Inn Road, London, England*
I. Dingwall-Fordyce,
*Department of Community Medicine, University of Aberdeen, Scotland*
*and*
C.F. Halliday,
*The Veterinary Invetigation Laboratory, Craibstone, Aberdeen, Scotland*

## Summary

*The potential of a computerised image analysis system for the study of interrelationships between environmental geochemistry and disease is shown by its ability to demonstrate a spatial association of copper deficiency in cattle with copper and molybdenum geochemistry and stream pH. Problems arising from small area statistics impaired a similar analysis of human data. These and related difficulties and their implications for the design of subsequent studies are outlined.*

## Introduction

"A direct relationship between geochemistry and human health is plausible, potentially exciting, repeatedly tantalising and to date rarely proven". When saying this, Crounse *et al.* (1983) described accurately the frustrations which are encountered in attempting to assess a potential interrelationship between human

---

[*] To whom correspondence should be addressed.

health and environmental agents. This relationship is even more difficult to assess when it is distorted by temporal and spatial factors, attenuated by the food chain and other extraneous factors, or displaced by population movement. The vast extent of exogenous factors which may contribute to human disease precludes their discussion here. Similarly, possibly important single genetic and polygenetic influences on interactions between man and the environment will not be discussed. Suffice to state that the disease pattern in any geographical area depends on the constellation of environmental factors and the susceptibility of individuals, and that any attempt to study the influence of geochemistry on human health obviously should take such factors into consideration. In this report we describe a preliminary application of a computer image analysis system to study the spatial distribution and interrelationships of geochemical and other environmental factors which may influence the development of disease in farm animals and man.

## Method and Background

The area for this study was the North East of Scotland. High quality geochemical data derived from analysis of stream sediments for approximately 30 elements have been prepared by the British Geological Survey (NERC). The area has a reasonably stable population of which 80% of population present in 1951 is still alive and resident in the region. The established interest, at the Rowett Research Institute, in the metabolism of minerals had enabled a local survey of the incidence of some deficiency diseases in animals and their correlation with local geochemistry. Further data on animal disease was collated by the local laboratory of the Scottish Veterinary Investigative Service. The data on human diseases were based on hospital discharge summaries which, since 1973 had been compiled and stored in the central computer of the Grampian Area Health Board.

Geologically the area contains a wide variety of rock types ranging from ultrabasic lithologies to highly evolved granites which outcrop over areas of tens of square kilometres, forming units of a size large enough for epidemiological study to be effective. There are also marked changes in geochemistry over the region, with high levels of Cr, Ni, Mg and of Mo, Fe, U, Pb, combined with exceptionally low levels of Cu, Ca and other essential trace elements in some areas. Hence the region provided an excellent opportunity to select "affected" and control groups in areas generally comparable but with contrasting geochemistries (Ashcroft et al., 1984). A system of shear belts comprising steep anastomosing, large and small scale shear zones representing discordant geological structures occur locally which are frequently associated with mineralising systems. Thus, for example, several discrete occurrences of molybdenum mineralisation occur which may be relevant to some aspects of disease in animals.

In the late 1960s the UK Institute of Geological Sciences (now the British Geological Survey) initiated a programme subsequently sponsored by the Department of Industry, to prepare geochemical maps of land areas of Britain. The sampling technique adopted was to colect stream sediments samples at an average density of 1 per $km^2$. Data are available for elements such as As, B, Ba, Be, Ca, Co,

Cr, Cu, Fe, K, La, Li, Mg, Mn, Mo, Ni, Pb, Rb, Se, Sn, Sr, Ti, U, V, Y, Zn, Zr. The geochemical data and geological information for the North East of Scotland were transferred to an International Imaging Systems ($I^2S$) image analysis system. Rainfall data for the area were supplied by the United Kingdom Metereological Office (Bracknell, Berkshire) for the years 1977, 1979 and 1981.

Population data of the study area on a 1 km$^2$ basis were supplied by the General Register Office (Scotland) (GRO(S)). The data received were small area statistic data from the 1981, 100% census of the United Kingdom. Preliminary plotting of these data revealed a major anomaly; many squares contained zero records for all classes of information, and some squares were missing from the data set entirely. These grid squares had been excluded for reasons of confidentiality. Such a restriction applies to all squares where the population is less than 25 people or where there are less than eight households. Two methods are applied to preserve confidentiality. The first is to suppress all information other than the total population, total number of males and total number of females, or the total number of householders respectively. Another device, "barnedisation", involves the application of a quasi-random number (which is either +1, 0 or -1) to all variables in the data set, providing the application of -1 does not make the new value negative; although it may reduce 1 to 0. This procedure will result in a minimal distortion to values in large areas or to values in small areas with a large population count, but it is obvious that the distortion is greater in small count areas or when the total count is broken down into smaller age and sex groupings. In such a situation there is a potential distortion ± 25%.

Another problem affecting reliability of the population data in this project is the postcode referencing system. Postcode areas are irregular in shape and variable in size, and are not always contained fully within a 1 km grid square. With the census data, population counts are attached to a postcode area centroid which does not always reflect the true geographical location of the population. However, the medical data are also referenced by postcode and since these were translated to grid references also, there was thought to be a measure of compatibility between population and medical data set referencing. To ensure complete compatibility in further studies computing algorithms for postcode to grid reference conversion should be examined.

Thus three problems exist in using small areas statistics data such as those involved in this study area where grid squares with less than 25 inhabitants represent a large proportion of the total; these are geographic location variability arising from postcode referencing, numerical distortion by "barnedisation", and suppression of data for reasons of confidentiality.

An alternative to small area statistics data was investigated. This was to analyse the 1981 census "selected population and household counts" data on a 1 km grid square basis. For each grid square this provided: the total population, total males and total females including visitors; the usually resident population for the total population categories; and total household counts. Despite its obvious limitations this data set was considered more accurately to represent the population pattern of the study area and it was used as the population data input for subsequent statistical processing of the medical data. The subset used was the "usually resident population" which excludes seasonal visitors to the area but takes into consideration male populations temporarily domiciled elsewhere as a consequence of their employment. The use of this data set as a population input precludes the analyses of morbidity data on the basis of the age and sex of individuals, and its use should only be regarded as an interim measure.

Another option available was the use of age/sex breakdowns for polygons (such as those which could be identified as geochemical "hotspots"). However the inherent problems of the small area statistics data also made this approach difficult.

Topographic data were derived in digital form from Ordnance Survey information on scale of 1:625,000. Woodland areas and urban centres were derived from Landsat Multispectral Scanner imagery. Thus river networks, road patterns, regional boundary and coastline information were digitised and incorporated as a separate overlay with British National Grid reference points.

The human diseases investigated were selected on the basis that hospital admission would be required for a definitive objective diagnosis. The diseases included were cancers of the oesophagus, stomach, colon, and rectum, ischaemic heart disease, cholelithiasis, and nephrolithiasis. Data on multiple sclerosis were incorporated from a continuing local epidemiological study.

For each morbidity data set the following information was available, sex of the patient, age on inclusion in the data base, postcode of the patient's address, with, additionally, for the multiple sclerosis data, addresses for patients' place of birth, childhood residence, and residence at presentation of the disease.

All the above data sets were gridded according to the National Grid reference into regions of 1 km grid squares. This provided a rectangular gridded region of 130 km x 80 km. In addition an "image" file was created for the NERC Computing Service (NCS) VAX 11/750 $I^2S$ image analysis system. This converted the 130 km x 80 km grid to a 512 x 512 picture element (pixcel) image without distorting the study region and permitted comparison with existing geochemical and geological data sets for the study area on the visual display unit.

The entire data base (Table 1) comprised over 300 files of information for each 1 km grid square. All data were geometrically rectified and registered to the British National Grid coordinance system.

---

**Table 1.** *Summary of data bases used in this study*

---

Topography
Geochemistry
Geology
Stream water pH and conductivity
Rainfall
Farm livestock: point occurrence of nutritional disease
Population census
Human disease

---

**Analysis of medical data**

Epidemiological data are not in general normally distributed and hence the choice of statistical method is more important if meaningful conclusions are to be obtained. The Poisson distribution closely follows the population observed in both the census and medical data sets for the data area and therefore was used in this project. The method developed for analyses of the data was dependent upon a comparison of the observed distribution of the human disease data within a 1 km grid cell with the computed expected disribution for that same cell. The probability of the observed value was calculated using the following formula

$$\text{Probability (POP)} = (EI^{AI}/AI)x - EI$$

where EI = expected incidence (standard morbidity rate, SM, of disease within area x population), AI = actual or observed incidence of disease, and POP = population of grid square. A separate programme was written to derive this transformation on a mainframe computer.

The probability value for each grid square in the study area may be interpreted as "the probability of finding the observed incidence of cases (AI) of the disease by chance, where the standard morbidity is SM, in a population of POP". The method gave a statistical measure of the observed incidence occurring by chance and was therefore a measure of the significance of the observed value.

The probability of disease occurrence was calculated for male and female populations for each grid square for the selected diseases. The calculations were computed on the NERC mainframe Honeywell computer. The results were subdivided into groups of significantly high and significantly low incidence (at the 1% level) and the values written to tape for transfer to the NERC $I^2S$ image processing system, where they were compared interactively with the other available data sets, notably geochemical data.

Since the probabilities were real numbers they could not be entered directly into the $I^2S$ system. To overcome this problem the log value of each probability was used and its exponent was entered into the $I^2S$ instead of the actual probability value.

Within the image processing system it is possible to overlay data base components rapidly. In addition the system is capable of performing Boolean arithmetic with the logical AND, OR NOT and XOR operations; hence, for example it is possible to instruct the system to process and display data bases using a masking technique to exclude extraneous data from areas of zero population or zero data thereby avoiding distortion of the upward data. This applied particularly to situations where areas of geochemical and medical data were different spatially, *i.e.* where disease data were available for areas with no mapped geochemical data or *vice versa*.

The $I^2S$ system is configured with eight graphics planes. Region of interest masks (regular or irregular in shape) can be drawn in these planes by manual interaction or density slicing of the image. In combination with the feedback capability of the system, masks can be used as information filters. A typical sample used in this study was "feedback to the display all copper geochemistry data for areas inside the mask in plane no. 5, where plane no. 5 contains a real distribution of significantly high incidence of oesophageal cancer". This operation could take two or three seconds to perform.

A further capability of the $I^2S$ image processing system which was used was the rapid computation of scattergrams to show the relationship between two variables. In this study the technique was used to look for possible correlations between selected diseases and certain geochemical trace elements.

## Appraisal of Data and Method using Nutritional Disease in Cattle

The area selected for geochemical investigations in this programme is noted for a long history of inorganic element deficiencies and toxicities in its population of cattle and sheep.

Manifestations of copper deficiency have been frequent and include, in young cattle, growth failure commencing during the first two years of life, greying of hair, and, in severe cases, anaemia. While the principal effects of copper deficiency in sheep are a variable incidence of neurological lesions leading to incoordination and death of lambs, other consequences can include a marked deterioration in wool structure and growth failure.

The geochemical background to this particular nutritional disorder in animals was investigated using the accumulated data sets since this would seem a particularly pertinent model with which to assess the methodology.

Copper deficiency in cattle is associated with the inhibitory effects of Mo upon Cu uptake by ruminants. The uptake of molybdenum from soils into edible crops such as pasture and other feeds is promoted by even modest increases in soil redox potential.

The compilation of data on deficiency disease distribution had to be confined to information derived from records of practising veterinarians and their supporting advisory laboratories. This identified 72 sites where clinical cases of confirmed Cu deficiency had occurred in the period 1974-1984. Since the animal population was confined principally to areas below the 1,000 foot contour, data from elevations above 1,000 feet were excluded using the "masking technique".

Since animal population statistics describing the total number of animals "at risk" in the sites of interest were not available the analysis of these data was restricted to a semiquantitative comparison the geographical distribution of reported outbreaks of copper deficiency and the distribution of possibly related geochemical anomalies.

No relationship existed between either the local geology or Cu geochemistry and the reported incidence of Cu deficiency and disease. However comparison of the incidence of Cu deficiency with Mo geochemistry displayed four areas of particular interest. In area 1 the high incidence of Cu deficiency was associated with anomalous Mo geochemistry. In area 2 the high incidence of Cu deficiency showed no clear association with high Mo geochemistry, additionally the data base for geochemistry was deficient in this area; in areas 3 and 4 there was an association between the occurrence of Cu deficiency and high Mo anomalies, especially when the effect of drift of top soil in relation to geochemistry is taken into consideration.

The association of copper deficiency with high Mo geochemistry is most tenuous in areas 1 and 2. However, when stream pH greater than 7 was displayed, a high proportion of the recorded instances of copper deficiency (52 from a total of 61) were found in localities where the stream pH exceeds 7. The exceptions were in areas 3 and 4 which have high Mo anomalies, and in area 2 where stream pH data were lacking along the coast.

Only three copper deficiency locations were associated with Mo geochemistry less than 1 ppm and of stream water pH less than 7, from 64 localities from which both Mo geochemistry and stream water pH data are available. This demonstrates a close relationship between Mo geochemisty, stream water pH, and Cu deficiency incidence. Despite the severe statistical limitations inherent in the data presented here the results are sufficiently interesting to merit further investigation in their own right, and, additionally, they illustrate the potential value of the analytical approach adopted in this study and demonstrate the value of being able simultaneously to investigate an interaction between more than two factors.

## Human Disease

In this account we will only discuss our examination of the data relating to carcinoma of the oesophagus and stomach, and carcinoma of the colon.

### Carcinoma of the oesophagus and stomach

The accuracy with which diagnoses are defined is a constant problem in medical data. For example, a tumour appearing at the junction between the oesophagus and the stomach could be identified as a lesion of either organ. Other problems arise from the fact that although we have based our study on conditions requiring hospital admission we have no indication about the incidence of undetected lesions occurring in the community.

The analyses of the human disease data was limited by the inability to standardise the morbidity ratios for the age and sex of the populations. As a consequence the Poisson distribution figures for the incidence of diseases such as carcinoma of the oesophagus probably reflected the age composition of the population in squares involved. Thus it showed a particularly low risk of occurrence in some of the newly populated areas of the region such as the recently developed suburbs of Aberdeen, whilst it showed increased risk in other areas in which populations have been more established and thus, on average, older or where there are nursing homes. A possible exception to this was the occurrence of cancer of the oesophagus in the area of the Spey Valley.

There are numerous geographical differences throughout the world in the incidence, prevalence and death rates for cancer of the oesophagus and stomach (Coggon and Acheson, 1984; Day, 1984). In the genesis of carcinoma of the stomach environmental factors from early childhood may contribute to this pattern (Correa et al., 1975).

Of particular interest has been the possible association of cancer of the oesophagus or stomach with soils with a low availability or content of Mo, and it has been suggested that this may be secondary to a requirement for Mo in microbial nitrogenase and nitrate reductase. Loss of activity of these molybdoenzymes in intestinal microflora may result in the persistence and formation of potentially carcinogenic nitrosamines. It has been postulated that these and related compounds exist in food and water and can additionally be formed in the stomach from nitrites and nitrates. Laboratory studies in rats, have shown that Mo inhibits the induction of tumours by the carcinogen N-nitrosocarcosine ethyl ester (Luo *et al.*, 1983). This hypothesis is far from proven but studies in rat models have shown a synergistic interaction between sodium nitrite and benzylethylamine and zinc deficiency in inducing papillomata and forestomach tumours (Fong *et al.*, 1984).

Several epidemiological studies have sought an association of geochemistry with the incidence or prevalence of carcinoma in the oesophagus and stomach. In the People's Republic of China, the availability of Zn and Mo in soils and their concentrations in local cereal crops correlated inversely with the rates of carcinoma of the oesophagus and stomach. It was suggested that associations with high soil pH, calcium carbonate content, and the intensive tilling of steep slopes and highly eroded soils over centuries led to low contents of Zn and Mo in soils. Similar factors and leaching similarly may have influenced adversely the Se content of the soils and, in turn, the Se intake of the indigenous population which may thereby experience an increased risk of carcinoma (Jackson *et al.*, 1986). The same authors have raised the possibility of an association of the lower availability of Zn from arid and humid soils and the raised incidence of carcinoma of the stomach in the Western and Northeastern states of the USA such as Wisconsin. They contrast this association with the local high availability of Se from such soils and the low death rate from heart disease (Jackson *et al.*, 1985).

In Transkei, Northeast Iran on the Caspian littoral, and Northern China oesophageal cancer is an important cause of death; disease incidences are respectively for males, 70, 115 and 130 per 100,000 population. Kibblewhite *et al.*, (1984) calculated incidence contours for oesophageal cancer in a district of Transkei. Foci of high and low incidences occurred within short distances (kilometres) of each other, and it was found that concentrations of Cu, Ni and B in the soils were lower in the high risk zones. The concentrations of Mn, Zn, and Mo were relatively low in the low incidence zones. An association between Mn deficiency and susceptibility to cancer of the oesophagus has also been postulated for the Finnish population (Marjanen *et al.*, 1972). However confirmatory studies are needed to substantiate these epidemiological data showing possible relationships between nutritional mineral imbalances and oesophageal cancer and need more supportive data.

We investigated the Poisson probability distributions for carcinoma of the oesophagus for the male and female populations of the study area. The limitations of these distributions have already been referred to. However, in Speyside, although the population is small, it has been calculated that the incidence of cancer of the oesophagus is approximately 175 per 100,000. However one has to be cautious about deriving any conlusions from these phenomena. Scattergrams were generated showing plots of the probability of the disease being abnormally frequent or infrequent against the geochemical data for Mo in general, Mo in areas with stream pH greater than 7, Zn, Se, U, As, and Cu. None of the scattergrams showed any apparent relationship between these factors. A similar analysis of cancer of the stomach was equally unrewarding.

## Cancer of the colon and rectum

It has been suggested that a low availability of environmental Se, or a low dietary intake of the element resulting in a reduced Se status of the population increases the risk of cancer in general, and that of the large intestine in particular (Willett *et al.*, 1981, Salonen *et al.*, 1984).

Separate probability plots for cancer of the rectum and colon in males and females were prepared. One important difficulty here is that if there is an already increased prevalence of the disease in the entire area under study then it may not be possible to detect appreciably higher incidences against this background. Some loci of apparently increased occurrence coincided with population centres. This phenomenon may well be related to the age distribution of the population and to the number of residential nursing homes situated in those areas, rather than to any discrete effect of an exogenous factor over and above these already prevalent in the region as a whole.

## Discussion

We have encountered a number of problems in this study. The principal difficulty was that relating to the use of the 1 km grid square and the severe restrictions which this placed upon the availability of the census data. As a consequence it was impossible to standardise our data accurately for age and sex. Clearly these are problems of small area statistics which would vex any attempt to look at rural populations in areas with a low population density.

Clearly, the selection of an appropriate scale in topographical analyses is important (Glick, 1982). Our choice of a fine resolution was to a large extent influenced by the data available and the size of the area under study. If a study were done for larger areas one could be able to use a lower resolution (say 10 km grid squares) which would not be subject to the limitations on human census data as were experienced in this project. However, had we used a grosser resolution of the data we would not have detected the relationship between geochemistry and Cu deficiency in cattle. Although our impression is that a study of a much larger area with a lower resolution would facilitate a more effective investigation, this level of investigation would present a considerable logistic effort. A suitable approach to make such an effort worthwhile would be to study a detailed continuous or discontinuous transcept of areas with contrasting geochemistries from a larger region such as Europe. This approach has been proposed for similar studies in the USA (Glick, 1982).

Additionally, studies of geochemistry and health would be more valuable if they included some indication of the body burdens of the elements under study. To achieve this one would need more accurate and reliable techniques than are currently available. However such an approach could include some measures of the trace element content of leucocytes, plasma, or red blood cells. For an ionic element such as iodine, measurements of urine have been used to assess relative iodine status and intakes of populations (European Thyroid Association, 1985).

It would be additionally naive to consider looking at interactions between health and geochemical environment in terms of single elements. This point is exemplified by our study of copper deficiency in cattle, and this preliminary study is encouraging in showing how effective the system used can be in highlighting areas where there is a risk of cattle or other livestock developing copper deficiency as a consequence of high environmental concentrations of an antagonistic element.

Our attempts to look at some human diseases in this study have been unsuccessful in detecting any clear associations. Our inability to standardise the data adequately for age and sex discouraged us from extensively exploring multiple interactions of geochemistry with the risks of diseases. Nevertheless this approach does provide a useful and rapid technique for inspection of relationships between data bases, masking of these to seek more definite interactions and to exclude extraneous or noncontributory information, and for the generation of scattergrams. The image analysis system however does not provide software for the rigorous statistical analysis of such derived data; in this respect the further application of the system to epidemiological studies would require development of additional software.

From the point of view of studying the association of geochemistry and health this system has however displayed its value; the study of copper deficiency in cattle has demonstrated that it will probably be effective in studying areas where populations are stable, dependent upon local subsistence crops, and where the period of residence in that area is continuous. This clearly could relate to less developed countries but it would still require accurate ascertainment of health statistics, and perhaps, some means of determining the population's intake and exposure to elements.

We have used cross-sectional retrospective data which are inherently suspect. Changing practices of acquiring and storing information over the past twenty years have limited our ability to analyse this data. It is clearly inappropriate to use 1981 census data to assess the standard morbidity rates for diseases occurring before that time; but earlier census data are recorded in different formats, have utilised different geopolitical divisions for representation, and have not been extensively computerised. Nevertheless analysis of such data can be used to create hypotheses and to devise new investigative strategies.

The multiple socio-economic, environmental and nutritional risk factors which may combine to influence health are difficult to untangle and it is not always clear whether such influences have an additive or multiplicative effect (Koopman 1981). There is a strong case for an extensive prospective epidemiological study of the diseases considered here in which geochemical and physical environments, lifestyles, culture and diets can be documented accurately. Ideally such a study would be strengthened if biological material were collected for later determination of relevant biochemical and genetic variables. This would require the compilation of many data bases on environmental and social factors. However some of these data are collected for other purposes such as mineral exploration, metereological monitoring, remote sensing of land use, population and socio-economic census data, and health statistics and may be available for epidemiological investigations. Additional information and the acquisition and preservation of biological samples could be a relatively straightforward and cost effective exercise. This type of exercise is an invaluable requirement for assessing the impact on human health of environmental factors such as geochemistry.

## Acknowledgements

This study was supported by the EEC and by the NERC. P J A thanks the Rank Prize Funds for additional support. Published by permission of the Director, British Geological Survey (NERC).

## References

Ashcroft, W.A., Kneller, B.C., Leslie, A.G. and Munro, M. (1984). *Nature*, 310, 760-762.

Coggon, D., Acheson, E.D. (1984). *Br. Med. Bull.*, 4, 335-341.

Correa, P., Haenszel, W., Cuello, C., Tattenbaum, S. and Archer, M. (1975). *Lancet*, 2, 58-60.

Crounse, R.G., Pories, W.J., Bray, J.T. an Mauger, R.L. (1983). Geochemistry and man: Health and Disease. In: I. Thornton (ed.), *Applied Environmental Geochemistry*, pp.267-308. Academic Press, London.

Day, N.E. (1984). *Br. Med. Bull.*, 40, 329-334.

European Thyroid Association; Sub-Committee for the Study of Endemic Goitre and Iodine Deficiency. (1985). *Lancet*, 2, 1289- 1293.

Fong, L.Y., Lee, J.S., Chan, W.C. and Newberne, P.M. (1984). *J. Natnl. Cancer Inst.*, 72, 419-425.

Glick, B.J. (1982). *Ann. Assoc. Amer. Geog.*, 72, 471-481.

Jackson, M.L., Zhang, J.Z., Li, C.S. and Martin, D.F. (1986). *Appl. Geochem.*, 1, 487-492.

Kibblewhite, M.E., Ban Rensburg. S.J., Laker, M.C. and Rose, E.F. (1984). *Environ. Res.*, 33, 370-378.

Koopman, J.S. (1981). *Am. J. Epidem.*, 113, 716-724.

Marjanen, H. (1972). *Ann. Agricultura. Fennae.*, 11, 391-406.

Salonen, J.T., Alfthan, G., Huttunen, J.K. and Puska, P. (1984). *Amer. J. Epidemiol.*, 120, 342-349.

Willet, W.C., Polk, B.F., Morris, J.S. *et al.*, (1983). *Lancet*, 2, 130-134.

# 9 Geochemical Environment Related to Human Endemic Fluorosis in China

Baoshan Zheng and Yetang Hong
*Institute of Geochemistry, Acadmia Sinica, Guiyang, Guizhou Province, P.R. China*

Possibly China is one of the countries most severely affected by endemic fluorosis in terms of both incidence and severity. It has been found that fluorosis occurs in every province and autonomous region of China. Fluorosis is found in 762 counties, accounting for nearly 36% of the total 2,136 counties in China. It is estimated that fluorosis patients may total more than 30 million.

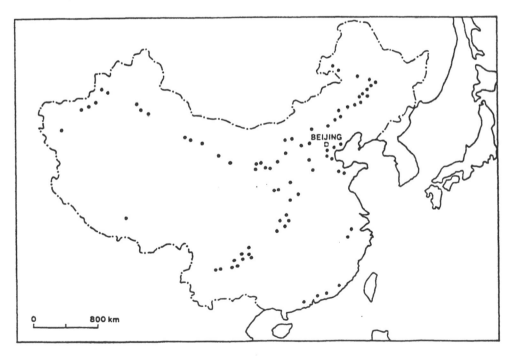

*Map showing the distribution of fluorosis districts in China.*

According to the sources of fluorine in the environment, the endemic fluorosis districts in China can be divided into five types associated with:

1. F-enriched phreatic waters;
2. pressure-confined waters with high F content;
3. F-enriched hot spring waters;
4. natural F-enriched foods; and
5. pollution by smoke as a result of burning F-rich coal.

## Fluorosis Districts with F-enriched Phreatic Waters

Fluorosis districts are most widespread, mainly in the arid or semi-arid regions such as Helongjiang, Jilin, Liaoning, Inner Mongolia, Hebei, Henan, Shanxi, Shaanxi, Ningxia, Gansu and Xinjiang provinces or autonomous regions. These districts constitute an endemic fluorosis-diseased belt which is situated on the periphery of the desert. The maximum F contents in surface and phreatic waters reach 129 mg/L and 40 mg/L, respectively.

An F-enriched geological background may be the geochemical explanation of the high F concentration in these districts. For example, in the Mt. Da Xinganling and Mt Yanshan areas are widespread Jurassic and Cretaceous rocks. These volcanic and intrusive rocks generally contain high F contents, ranging from 500 to 800 ppm. And in the areas of Shaanxi, Ningxia and Gansu provinces are widespread loess deposits of great thickness. According to the determinaton of F in our laboratory, the total F content of this loess ranges from 490 to 550 ppm (Liu Dongsheng et al., 1980).

In addition, it seems to be even more important that soda salinisation is well developed in soils within this geological background. The districts where soils have been affected by soda salinisation are usually fluorosis ones (Zheng Baoshan, 1983).

## Fluorosis Districts with Pressure-confined Waters of High F Concentration

These districts are distributed mainly in the region around the Bohai Gulf, where semi-arid, alternatively marine-terrestrial strata are exposed with a high salinity.

Tianjin City, which is the third biggest city of China, is located in this area. The F content of deep well water in the suburbs of Tianjin City is 7 mg/L. Patients seriously affected by fluorosis have been observed in the suburbs.

Cangzhou City of Hebei Province has 400,000 residents who drink pressure-confined waters with an F concentration of 3-8 mg/L.

The reason why the F concentration in such pressure-confined waters is so high still remains open to question.

## Fluorosis districts with F-enriched hot spring waters

There are 2,493 hot springs in China. They are distributed mainly in Yunnan, Guangdong, Fujian and Taiwan provinces. Most of these hot springs have higher F concentrations. Fluorosis occurs in the areas around some of the well-known hot springs, such as Xiaotangshang in Beijing, Lintong in Shaanxi province, Tangkeng in Guangdong province and Longxi in Fujian province.

In Kaga village of Xietongmen county of Tibet, there is a hot spring with a temperature of 60°C. The F content of the hot spring water reaches 15 mg/L. The residents living in the three villages around the hot spring have to drink such high F water.

### Fluorosis districts with natural F enriched foods

It is well-known that tea-leaves generally have a high F content. It is also clear that drinking tea does not usually lead to fluorosis.

Drinking extremely rough tea, however, is considered to be a reason for fluorosis occurrence with some Tibetan herdsmen in Tibet and Qinghai autonomous regions and Sichuan province, because these areas are located on the plateau at 40,000 m above sea level. In these areas both fresh vegetables and fruits are lacking in the diet. Main foods are highland barley, animal meat products and dairy products. The herdsmen are used to drinking a lot of rough tea to help to digest these foods.

### Fluorosis Districts with Smoke Pollution as a Result of Burning Coal Indoors

According to its seriousness, this type of fluorosis comes next in China. It is mainly distributed in the south of China including Hubei, Hunan, Guizhou, Sichuan, Yunnan provinces and so on.

Endemic fluorosis in Zhijin county and Guizhou province can be taken as an example. The county is situated in the south western part of Chinan. Its altitude is about 1,400 m above sea level. There is a village named Hehua which is located about 10 km west of Zhijin township. It is a severe fluorosis district. The residents in Hehua village, like most other districts affected by this type of fluorosis, are minority people, mainly the Miao people. The drinking water source for local residents is surface water or shallow well water. It is interesting that the average content of F in drinking water from this area is about 0.15 mg/L, which is far below that of the standard drinking water. So, in this case fluorides in drinking water sources do not seem to be responsible for the fluorosis the local residents are suffering (Anti-epidemic Station of Shien District, Hubei Prov., 1980; Zheng Baoshan *et al.*, 1984).

The main food crop in Hehua village is corn. The F content of fresh corn is also lower than the average F content of 0.4 µg/g.

In this area, however, there is a moist climate. The annual rainfall is around 1,000 - 1,500 mm, and the humidity is high. Fortunately in the area of Zhijin county occurs a stratified coal system consisting of sandstone, marl, clastic rocks and anthracite. In order to prevent fresh corn from mildewing and going rotten, the local residents usually bake corn over a coal fire after harvest. Generally, corn is put on a bamboo shelf or hung under the ceiling above a stove without a chimney.

Due to contamination by coal smoke and respiration of the fresh corn, the F content of baked corn ranges from 10 to 100 µg/g with an average of 41 µg/g,

95

depending on the coal type, the moisture content of fresh corn and the length of baking time.

According to the determinations of 138 anthracite samples in our laboratory, the F content ranges from 80 to 2,000 ppm with an average of 220 ppm in contrast to the global average of 80 ppm in coal as suggested by Bown (1979).

Besides corn, other vegetables, such as hot pepper, pumpkin, *etc.*, are often contaminated by coal smoke. It has been shown that the average F content of baked pepper amounts to 460 $\mu$g/g.

It has been estimated by some Chinese medical units that the total amount of F ingested by the local residents in this area is about 18 mg/day, of which about 63% comes from food, especially corn.

As a result, the local residents of Hehua village suffer serious fluorosis diseases. According to the data from a health check of 100 schoolchildren, 94 children suffered from skeletal fluorosis of varying degrees.

In recent years the Chinese government has initiated propaganda about the harm caused by the use of coal fires without chimneys indoors. Local governments at various levels help the local residents of diseased areas rebuild and improve their cooking stoves. It may be said that the occurrence of this type of fluorosis is probably related to some interactions. Most of the areas affected by this type of fluorosis are either mountainous areas or minority people-inhabited areas with relatively poor nutrition and sanitation. The moist climate and the extensive occurrence of coal are the two initial reasons why fresh corn has to be baked over coal fires. F released during the burning of anthracite high in F will directly contaminate food. However, the influence of polluted air and dust as well as other elements released from coal still awaits further investigation.

This type of fluorosis presents a case in which F in the geological environment affects human health through contamination of food and air. In this situation, F released from coal is the cause of fluorosis rather than natural fluoride-high waters.

## References

Anti-epidemic Station of Shien District, Hubei Province (1980). *Chinese J. Prev. Med.*, **14**(3), 164.

Liu Dongsheng *et al.* (1980). *Geochimica*, **1**, 13-22.

Zheng Baoshan (1983). *Acta Scientiae Circumstantiae*, 3(2), 113- 122.

Zheng Baoshan (1984). Unpublished paper.

# 10 Environmental Geochemistry of Trace Elements in Poland

Alina Kabata-Pendias
*Institute of Soil Science and Plant Cultivation, 24-100 Pulawy, Poland*
*and*
Róza Uminska
*National Institute of Rural Medicine, 20-950 Lublin, Poland*

## Summary

*Anthropogenic fluxes of trace elements into the light sandy, acid soils, which predominate in Poland, have significant ecological effects. The most important health risk is related, at the national level, to increased contents of Cd and Pb in various environmental compartments.*

## Introduction

Geological structures and climatic features of the landscape in Poland have formed mainly light sandy soils, which are quite frequently podsolised and acidified. Thus, sandy acid soils form about 50% of arable land.

Geochemical and mineralogical properties of the soils have created favourable conditions for deficiencies of some micronutrients. Geochemical provinces of iodine and cobalt deficiencies in humans and animals were broadly investigated, whereas deficiencies of boron, copper and zinc were observed in various crops. Recently, the deficiency of selenium in domestic animals has also been recognised in certain regions of Poland.

During the period 1950-1970 studies on trace element distribution in soils, their behaviour and phytoavailability were carried out. Dietary investigations were also broadly developed. These studies have resulted in the estimation of background contents of trace elements in soils and plants (Tables 1 and 2), as well as in dietary guidelines.

Recent rapid development of industry based on energy from coal combustion has created a chemical stress on the environment in Poland. Trace inorganic pollutants have become targets of most of the studies carried out during the last decade in Poland.

97

**Table 1.** *Trace elements in surface soils of Poland (values commonly found) (mg/kg DW).*

| Element | Podsols | Combisols | Histosols |
|---------|---------|-----------|-----------|
| As | 0.8 | 5 | 13 |
| B | 10 | 30 | 25 |
| Be | 3 | 15 | 0.5 |
| Cd | 0.1 | 0.3 | 0.05 |
| Co | 5 | 8 | 3.0 |
| Cr | 40 | 60 | 15 |
| Cu | 6 | 15 | 5 |
| F | 70 | 250 | 150 |
| Hg | 0.05 | 0.2 | 0.02 |
| Mn | 300 | 600 | 150 |
| Mo | 1.5 | 3 | 1 |
| Ni | 8.0 | 15 | 10 |
| Pb | 20 | 30 | 25 |
| Se | 0.1 | 0.3 | 0.34 |
| V | 15 | 45 | 60 |
| Zn | 25 | 90 | 60 |

**Table 2.** *Trace elements in cultivated plants in Poland (values commonly found) (mg/kg DW).*

| Element | Cereals | Grasses grain | straw | Leafy Vegetables |
|---------|---------|--------|-------|------------------|
| As | 0.2 | 0.4 | 0.8 | 0.5 |
| B | 0.8 | 5.0 | 10 | 30 |
| Cd | 0.2 | 0.1 | 0.2 | 0.2 |
| Co | 0.2 | 0.1 | 0.2 | 0.4 |
| Cr | 0.02 | 2.5 | 2.0 | 10 |
| Cu | 4 | 3 | 6.0 | 15 |
| F | 4 | 12 | 7.0 | 8 |
| Hg | 0.01 | 0.05 | 0.02 | 0.2 |
| Mo | 0.4 | 0.6 | 2.5 | 1.5 |
| Mn | 25 | 45 | 150 | 30 |
| Ni | 0.5 | 0.7 | 0.9 | 3.5 |
| Pb | 0.3 | 3 | 5 | 12 |
| Se | 0.02 | 0.03 | 0.1 | 0.1 |
| V | 0.08 | 0.9 | 2.0 | 0.8 |
| Zn | 35 | 40 | 40 | 100 |

**Table 3.** *Emission of cadmium in some European countries (1979).*

| Country | Total emission ($10^6$ g/year) | Percent of total emission in Europe | Relative ratio of emission to: | |
|---|---|---|---|---|
| | | | population (g/capita) | territory (g/ha) |
| Belgium | 171 | 6.2 | 17 | 5 |
| France | 170 | 6.2 | 3 | 3 |
| FRG | 328 | 12.1 | 6 | 13 |
| Poland | 207 | 7.6 | 6 | 6 |
| USSR | 816 | 30.2 | 4 | 1 |

After: Pacyna (1986)

**Table 4.** *Fluxes of trace metals from the atmosphere to land surface in Poland (1981-1983) (g/ha/year).*

| Metal | Rural | Region Little industrialisation | Heavy industrialisation |
|---|---|---|---|
| As | 1.7 | 10.3 | - |
| Cd | 2.3 | 3.1 | 4 |
| Cu | 29 | 53 | 257 |
| Hg | 0.3 | 0.8 | - |
| Mn | 154 | 413 | 804 |
| Pb | 104 | 181 | 460 |
| Zn | 358 | 516 | 989 |

After: Kabata-Pendias (1985), Manecki *et al.* (1981).

## Environmental Compartments

*Air*

Most trace elements of industrial origin are released to the atmosphere. Calculations by Pacyna (1986) of emissions of trace metals in Europe indicate that Poland contributes quite a significant proportion of their total emission. Annual emission of cadmium in Poland is fairly similar to the amount of this metal emitted in other European countries (Table 3). Similar proportions have been calculated for other

**Table 5.** *Trace elements in wastes proposed for soil amendment (mg/kg DW) in Poland.*

| Element | Sewage sludges | Bottom ash after lignite combustion | Waste lime |
|---------|----------------|-------------------------------------|------------|
| As | 2-60 | 3-30 | 1-1250 |
| B | 15-1,000 | 90-1,000 | |
| Cr | 20-40,600 | 10-50 | 9-13,000 |
| Sn | 40-700 | - | |
| Zn | 700-49,000 | 15-1,000 | |
| F | 2-740 | - | |
| Cd | 2-1,500 | 0.7-7.5 | 1-474 |
| Co | 2-260 | 7-17 | |
| Mn | 60-4,000 | 67-3,171 | |
| Cu | 50-3,300 | 8-500 | |
| Mo | 1-40 | - | |
| Ni | 16-5,300 | 28-53 | |
| Pb | 50-3,000 | 12-30 | 15-30,250 |
| Hg | 0.1-55 | - | |
| Se | 2-9 | - | |
| Sr | 40-360 | - | |
| U | - | 1-3 | |
| V | 20-400 | 1-2 | |

trace metals. Measurements of the atmospheric input of cadmium and lead to surface soils also give the comparable values for Poland and other countries in Europe (Kabata-Pendias and Pendias, 1984).

Spatial distribution of trace element emissions in Poland was not estimated. However, this can be easily related to the localisation of industrial plants and industrial waste production, which is concentrated in the south-western region of the country. There are no available data on the long-distance transport of trace inorganic pollutants, however, a relatively high fallout of these metals in rural areas of the country clearly indicates their aerial fluxes (Table 4).

*Soils*

Soils of all terrestrial ecosystems are most often the final sinks of anthropogenic trace pollutants. Thus the pollution of surface soils with trace elements is quite common in industrial regions. Also soils of orchards and gardens are known to be already contaminated due to the utilisation of fertilisers and various wastes (Table 5).

Waste lime as a source of the trace metals is of special concern in agriculture. Highly acidified arable soils have a great need for liming, and therefore several waste

**Table 6.** *Trace elements in surface contaminated soils (mg/kg DW) in Poland.*

| Element | Maximum value found | Source of pollution |
|---------|---------------------|---------------------|
| Cd | 290 | Metal smelter |
|    | 5 | Gardening |
| Be | 50 | Metal smelter |
| Cu | 1,200 | Metal smelter |
|    | 240 | Gardening |
| Hg | 0.2 | Seed treatment |
| Pb | 4,650 | Metal smelter |
|    | 165 | Gardening |
| Zn | 10,000 | Metal smelter |
|    | 100 | Gardening |
| F | 13,200 | Al-smelter |

limes, (especially after metal ore flotation) are widely utilised (Table 5). Studies on the impact of waste limes on trace metals in arable soils show that cadmium and lead are most likely to be increased in surface soils (Uminska, 1985; Uminska and Kobylska, 1985).

Soils of industrial regions have been intensively studied and several reports have described local situations of trace elements. The highest reported concentrations of trace elements, as shown in Table 6 exceed several times their background content and are likely to be of an ecological significance.

Mobility and phytoavailability of anthropogenic trace metals in soils are rather high due to acidity, low content of organic matter, and sandy structure. Thus, a high rate of metal leaching with seepage waters should be expected in this kind of soil. Actually, light acid soils are not long-lasting sinks of trace elements. Especially in the case of anionic elements, like fluorine or boron, pollution of ground waters is observed due to leaching processes.

*Waters*
Rivers and lakes are known collectors of a significant proportion of the anthropogenic trace elements. Their cationic forms are known to be deposited rapidly in bottom sediments. Nevertheless, in river water close to the input of industrial wastes, an increased level of soluble metals is observed (Table 7). Enrichment factors for the metals are very high, when compared to unpolluted rivers of rural regions. Increased levels of trace metals in terrestrial waters of Poland are known to contribute to their transfer to the human food chain by zoo- and phyto- planktons and by fish and water plants.

**Table 7.** *Trace metals soluble in water of the upper part of Wisla river* (μg/L).

| Metal | Range of concentration | Pollution factor[*] |
|-------|------------------------|------------------|
| Cd | 0.5-7 | 35 |
| Cr | 3-300 | 51 |
| Cu | 10-300 | 50 |
| Fe | 80-17,600 | 16 |
| Mn | 30-1,200 | 75 |
| Ni | 10-300 | 60 |
| Pb | 5-120 | 48 |
| Zn | 16-300 | 21 |

After: Helios-Rybicka (1986).
[*] Ratio of maximum metal concentration in water to its background value for rivers in rural areas.

*Plants*

Two pathways of trace elements to plants, root uptake and foliar uptake, are important in the polluted environment. In general, chemical composition of plants reflects both the elemental composition of the growth media and the atmospheric input of pollutants. In several regions of Poland, however, root uptake seems to be an important flux of the metals into the food chain.

Several baseline surveys were recently carried out on trace element status in plants. The projects were focused mainly on food and fodder crops, and on industrial sites especially. Some of these projects were conducted for several years and yielded background data for various environmental compartments (Tables: 1, 2, 5, 6, 8).

Vegetation in industrial areas is heavily polluted with the metals of which cadmium shows the highest rate of enrichment (Table 8). Wheat grain used as an indicator for the environmental pollution contains relatively stable amounts of most of the metals (Table 9). The only significant enrichment was observed for cadmium and zinc in grain collected from the heavily industrialised region. Also copper and iron were slightly increased in grain of the industrial area.

**Environmental Health Aspects**

A close relationship between the geochemistry of the environment and animal and human health has already been recognised. In each case study, however, a simple relationship is not easily observed.

Results of several studies conducted recently in Poland indicate significant increases of cadmium and lead in various environmental compartments (Amarowicz

**Table 8.** *Heavy metals and fluorine contamination of plants grown in industrial regions of Poland.*

| Element | Plant (tops) | Content (ppm DW) |
|---------|--------------|------------------|
| Cd | Lettuce | 5-14 |
| Cu | Dandelion | 73-274 |
| F | Clover | 14-173 |
| Ni | Clover | 3-15 |
| Pb | Lettuce | 45-69 |
| Zn | Lettuce | 213-393 |

**Table 9.** *Trace elements in wheat grain in various regions of Poland (mg/kg DW) (average values for three years collection 1981-1984, n = 127).*

| Element | Regions variously industrialised | | | C/A |
|---------|-------------|-------------|-------------|-----|
|         | Little A | Medium B | Heavily C | |
| B | 0.8 | 0.6 | 0.9 | 1.1 |
| Cd | 0.1 | 0.1 | 0.5 | 5.0 |
| Cr | 0.3 | 0.1 | 0.1 | 0.3 |
| Cu | 2.1 | 4.8 | 3.2 | 1.5 |
| Fe | 24 | 30 | 39.5 | 1.6 |
| Mn | 21 | 22.5 | 22 | 1.0 |
| Ni | 0.6 | 0.3 | 0.3 | 0.5 |
| Pb | 0.4 | 0.3 | 0.3 | 0.8 |
| Zn | 21 | 31.5 | 51 | 2.4 |

*et al.*, 1985; Gzyl *et al.*, 1984; Marchwińska *et al.*, 1984; Marzec and Bulinski, 1985, 1986; Nabrzyski and Gajewska, 1984; Zommer-Urbańska and Topolewski, 1984). It has been shown that children (up to 4 years old) are especially exposed to increased levels of lead, which is observed in an elevated lead content of hair (infants - 9.5 mg/kg dry hair, and children - 6.9 mg/kg dry hair: Skorkowska-Zieleniewska *et al.*, 1984). The total population of Poland seems to contain a slightly higher average cadmium level in blood (1.04 µg/100 mL) than people in the United States of America (0.85 µg/100 mL) (Buliński *et al.*, 1982). Calculated daily intake of these two metals shows that children, in the extreme cases, can absorb more cadmium and

**Table 10.** *Daily intake of cadmium and lead by children and adults in Poland* (μg/day).

| Metal | Uptake by: | |
| | Children (15 kg body weight) | Adults (60 kg body weight) |
|---|---|---|
| Cadmium | | |
| Lowest | 15 | 20 |
| Highest | 300 | 150 |
| | | |
| Lead | | |
| Lowest | 40 | 100 |
| Highest | 1000 | 500 |

After: Szteke and Jedrzejczak (1986).

lead than adults (Table 10). The highest daily uptake of the metals by adults exceeds DAI (Daily Allowable Intake) values given by WHO. Detailed investigations of the pathways of cadmium and lead in human diets reveal that the main source of these metals is related to plant consumption which contributes about 70% of metals in the total diet.

## Conclusions

Current studies on environmental geochemistry in Poland have been focused on environmental health aspects. Local and regional problems are closely related to the sources of trace inorganic pollutants. In the whole country however, cadmium and lead appear to be a significant health risk.

Present needs for further national research are related to the following problems:
- guidelines and legislation for trace element levels in soils,
- maximum permissible concentration of trace elements in food (dietary guidelines),
- special protection of children from exposure to lead and cadmium,
- fluxes of trace elements between environmental compartments under anthropogenic stress,
- environmental and technical barriers protecting humans against exposure to trace inorganic pollutants.

# References

Amarowicz, R., Smoczynski, S. and Markiewicz, K. (1985). Chemical contamination of daily food rations in selected groups of population. *Roczniki PHZ*, **36**(6), 461-466.[*]

Buliński, R., Bloniarz, J., Dabrowska, D., Koktysz, N., Kot, A., Marzec, Z. and Szydlowska, E. (1982). Study on tissue cadmium content in general population individuals, and in those employed under heavy work conditions and exposed to environmental contamination in the province of Lublin, *Bromat. Chem. Toksykol.*, **15**(3), 231-233.[*]

Gzyl, J., Kucharski, R. and Marchwińska, E. (1984). Evaluation of lead exposure in inhabitants of certain areas of the Legnica- Glogów copper mining district by the indicator method. *Roczniki PZH*, **35**(5), 399-403.[*]

Helios-Rybicka, E. (1986). Fixation of heavy metals by clay minerals in bottom sediments of the upper part of Wisla river. Zesz. Nauk. AGH, *Geologia* 32, 123.[*]

Kabata-Pendias, A. (1985). Chemical composition of atmospheric particles in Pulawy area. *Studia Pulawskie, Seria B*, **1**, 33-52.[*]

Kabata-Pendias, A. and Pendias, H. (1984). *Trace Elements in Soils and Plants*, pp.315. CRC Press, Inc. Boca Raton, Florida.

Manecki, A., Klapyta, Z., Schejbal-Chwastek, M., Skowronski, A., Tarkowski, J. and Tokarz, M. (1981). Impact of industial pollution on environmental geochemistry of Puszcza Niepolomicka. Prace min. 71, pp.1-58. PAN o/Kroków.[*]

Marchwińska, E., Kucharski, R. and Gzul J. (1984). Cadmium and lead concentrations in samples of potatoes from various regions of Poland. *Roczniki PZH*, **35**(2), 113-118.[*]

Marzec, Z. and Buliński, R. (1985). Study on some trace elements in home-made foodstuffs. Part, V Chromium, nickel, and selenium content in vegetable and fruit preserves. *Bromat. Chem. Toksykol.*, **18**(1), 17-20.[*]

Marzec, Z. and Buliński, R. (1986). Studies on some trace elements content in foodstuffs of home growth. Part IX. Chromium, nickel, and selenium content in mushrooms and fruits. *Bromat. Chem. Toksykol.*, **19**(2), 84-87.[*]

Nabrzyski, M. and Gajewska, R. (1984). Determination of mercury, cadmium and lead in food. *Roczniki PZH*, **35**(1), 1-11.[*]

Pacyna, J.M. (1986). Atmospheric trace elements from natural and anthropogenic sources. In: J.O. Nriagu and C.I. Davidson (eds), *Toxic metals in the atmosphere*. pp.33-52. J. Wiley, London.

Skorkowska-Zieleniewska, J., Cabalska, B., Golabek, B., Symonowicz, H. and Nowakowska, M. (1984). Biochemical assessment of lead content in hair in a mixed population. Part II. *Roczniki PHZ*, **35**(4), 337-340.[*]

Szteke, B. and Jędrzejczak, R. (1986). Assessment of cadmium and lead pollution of food in Poland. Manuscript in Polish.

---

[*]Published in Polish.

Umińska, R. (1985). Evaluation of level of selected heavy metals in Polish soils - a contribution to environmental exposition. *Proceedings of the 1st International Symposium on Geochemistry and Health*, pp.68-71. London.

Umińska, R. and Kobylska, B. (1985). Evaluation of level of arsenium, chromium, cadmium and lead in the soils amended with industrial wastes. *Prace Nauk. Polit. Szczecinskiej*, **291**, 95-99.[*]

Zommer-Urbańska, S. and Topolewski, P. (1984). Fluorine content in vegetables and fruits cultivated in the region of fluorine compounds emission by glass works "Irena", *Inowroclaw. Bromat. Chem. Toksykol.*, **17**, 153.[*]

# 11 Geochemistry and Health in Sri Lanka

C.B. Dissanayake[*],
*Department of Geology, University of Peradeniya, Sri Lanka and Institute of Fundamental Studies, Kandy, Sri Lanka*
M.U. Jayasekera,
*Institute of Fundamental Studies, Kandy, Sri Lanka*
*and*
S.V.R. Weerasooriya,
*Institute of Fundamental Studies, Kandy, Sri Lanka*

## Summary

*In view of the fact that Sri Lanka has clearly demarcated climatic zones, physiographic divisions, soil groups and geological formations, the geochemical distribution of chemical elements and their impact on human health provide interesting study. The fact that the vast majority of the people of Sri Lanka live in very close contact with the immediate physical environment enhances the usefulness of epidemiological studies in Sri Lanka.*

*Large areas in the Dry Zone of Sri Lanka, particularly the mineralised terrains with abundant geological and structural discontinuities have high concentrations of fluorides. The incidence of dental fluorosis is high in these areas and in cases where the fluoride concentrations in the drinking water reached levels as high as 10 ppm, mottling of teeth among school children was very prominent. It has been found that with the recent introduction of deep wells to the Dry Zone, the water from these wells, in many cases, contained excessive amounts of fluorides.*

*Human cancer, particularly oesophageal cancer is clearly geographically distributed and correlates well with the nitrate levels in the groundwater. Nitrates are abundant in areas of high population density, high nitrogenous fertiliser use and in general in the wet zone of Sri Lanka.*

*Endemic goitre is also widely prevalent in Sri Lanka with an estimated 10 million people at risk. Even though iodine deficiency is the chief cause, other factors such as the abundance of Co and Mn are also important etiologically.*

---

[*] Author to whom correspondence should be addressed.

# Introduction

The application of geochemistry to problems of health and environmental studies has been carried out only very recently in Sri Lanka. With the establishment of the first Geology Department at the University of Peradeniya in 1967, Geochemistry as a scientific discipline was first introduced into the university curriculum. With the establishment of the Environmental Geochemistry Research Group (EGRG) of the Department of Geology, University of Peradeniya, headed by the senior author, research into many aspects of the environment of Sri Lanka commenced (Dissanayake and Hapugaskumbura, 1980; Dissanayake and Ariyaratne, 1980; Dissanayake and Jayatilaka, 1980; Weerasooriya et al., 1982; Dissanayake et al., 1984; Dissanayake, 1984). The establishment of the EGRG led to the geochemical investigation of city canals, rivers, polluting environments, natural and man-made lakes, rain and tap water, etc. and first hand information on some aspects of the status of the environment of Sri Lanka was obtained. By this time, other organisations such as the Central Environmental Authority of Sri Lanka, Departments of Chemistry and Community Medicine in many universities, Water Supply and Drainage Board, National Aquatic Resources Agency (NARA), Division of Occupational Hygiene, Ministry of Labour, Ceylon Institute of Scientific and Industrial Research (CISIR), had embarked on research projects dealing with the environment. The application of geochemistry to health has only just been initiated and there seem to be vast possibilities for the geochemist to make extremely useful contributions to epidemiological studies in Sri Lanka. Sri Lanka offers an ideal opportunity to the geochemist to make extremely useful contributions to epidemiological studies in Sri Lanka. Sri Lanka offers an ideal opportunity to the geochemist in view of the fact that the vast majority of the people are intimately associated with the physical environment with less than 10% having access to piped water. The geochemistry of the physical environment governs to a very great extent, the general health of these people and geographic distributions of certain diseases are clearly observed. Further, out of the 10 great soil groups, 9 are found in Sri Lanka and this affords an opportunity to the geochemist to correlate health with soil chemistry.

The recent compilation of the first Hydrogeochemical Atlas of Sri Lanka (Dissanayake and Weerasooriya, 1985) has provided much new information that could be effectively used for epidemiological studies in Sri Lanka. The "Soil, Vegetation and Health Group" of the newly established Institute of Fundamental Studies has as its objectives the application of geochemistry in studies on dental diseases, goitre and cancer. This paper deals with the recent advances made in Sri Lanka in studies pertaining to the applicaton of geochemistry in health.

# The Geology and Geomorphology of Sri Lanka

Geologically the greater part (about 92%) of the country consists of rocks of Precambrian age, the island having remained stable over a long period of time. The Precambrian rocks of Sri Lanka are identified as forming two major divisions, namely the Highland Group and the Vijayan Complex (Cooray, 1978). The Highland Group has a metamorphic subdivision termed the Southwest Group. The Vijayan

Complex is geographically separated into the western and eastern Vijayan Complex by the linear arcuate fold belt of the Highland Group (Figure 1).

A suite of metasedimentary and metavolcanic rocks formed under granulite facies conditions comprises the Highland Group. Among the metasediments, quartzites, marbles, quartzo-feldspathic gneisses and metapelites form the major constituents. The Southwest Group consists mainly of calciphyres, charnockites and cordierite-bearing gneisses. The western Vijayan rocks on the other hand consist of basement type leucocratic biotite gneisses, migmatites, pink granitic gneisses and granitoids with compositions varying from granitic, syenitic to granodioritic. The granitoids frequently have enclaves of amphibolite and hornblende gneiss. Bodies of metasediments are rare in western Vijayan. The eastern Vijayan is composed of biotite hornblende gneisses, granitic gneisses and scattered bands of metasediments and charnockitic gneisses. Small plutons of granites and acid charnockites also occur close to the east coast (Corray, 1978). The chemical composition of these rocks has a direct bearing on the geochemistry of the water and, as will be shown later, on the general health of the community.

Sri Lanka, with an area of 69,450 $km^2$, is primarily a part of the shield area which comprises peninsular India. Geologically and physically, Sri Lanka is a southern continuation of India, being only recently separated from the mainland by the shallow sea covering Palk Strait and the Gulf of Mannar. On the basis of height and slope characteristics, the island can be divided into 3 main morphological regions (Vitanage, 1970) (Figure 1).

1. The coastal lowlands with elevations from sea level to 300 m with a few inselbergs. Slopes are generally flat lying in the narrow marshy belt along the coastal fringe while further inland low "turtle backs" appear.

2. Uplands with elevations from 300 m to 915 m consisting of ridge and valley topography and highly dissected plateaus with narrow arenas and domes occupying nearly 30% of the island.

3. Highlands with a series of well defined high plains and plateaus rimmed with mountain peaks and ridges with elevations greater than 915 m, characterise the central part of Sri Lanka. High level topographic discontinuities are common and these form the boundary of a series of high plains, plateaus and structural terraces. Laterites and laterisation are common in these places.

Sri Lanka which has a typical humid tropical climate lies in the monsoon region of south east Asia. The island is characterised by clearly demarkated dry and wet zones as shown in Figure 1. The average mean temperature of the wet zone lies between 70-85°F and in the dry zone it is approximately 90°F. Depending on the altitude, the mean temperatures of the Highlands vary betweeen 58°F and 78°F.

## The Hydrogeochemistry of Groundwater of Sri Lanka

Figure 2 illustrates the distribution of the major water types in Sri Lanka, namely: (a) Calcium type, (b) Magnesium type, (c) Sodium/potassium type, (d) non-dominant cation type.

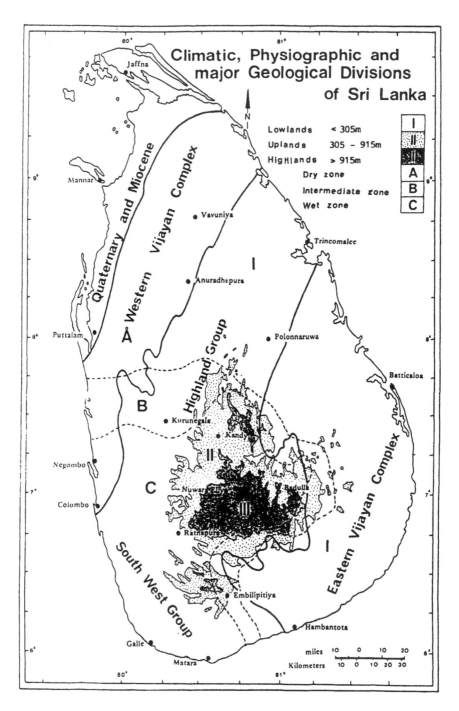

**Figure 1.** *Map showing the main physical divisions of Sri Lanka.*

*The calcium type*

In Sri Lanka, this type of water is distributed mainly in the northern, central and in some parts of southern, eastern and north central regions. The Cl type predominates in the northern parts whereas the $HCO_3$ type is prevalent in the central regions. The effect of salinity and the presence of carbonate rocks in these areas could possibly be attributed to such a distribution. The total dissolved solids (TDS) show significant correlations with K, Ca, $HCO_3$ and Cl. The transition elements however do not show significant correlations for this type of water.

*The magnesium type*

When compared to the other types of water, the magnesium type is distributed in relatively smaller areas, the southern part of the country around Embilipitiya having higher concentrations. In this type of water, only the Cl and $SO_4$ sub-types could be found.

*The sodium/potassium type*

This type forms a major group and is distributed widely in Sri Lanka, particularly around the central region. The north western and north central and the south eastern dry zones mainly contain this type of groundwater. From among the sub-types, the Cl type is predominantly found in these regions. Excessive evaporation and the probable influence of salinity may have contributed to the prevalence of this water type.

*The non-dominant cation type*

As illustrated in Figure 2 the non-dominant cation type of water is distributed mainly at the periphery of the central Highlands and in some parts of the north central and southern regions. The $HCO_3$ and non-dominant anion sub-types predominate in these regions.

*Effect of geology and climate on the chemistry of groundwater*

A closer study of the distribution patterns of the groundwater types in Sri Lanka reveals that the underlying geology and the climate affect the chemical quality of water to a great extent. The wet zone of Sri Lanka consists for the most part of non-dominant cation types, Ca-$HCO_3$ and non-dominant anion types. In the dry zone however, the Na/K type predominates and in this type of water the Cl sub-type is found covering vast areas of the dry zone. Evaporation under the strong drought conditions which prevail in the dry zone of Sri Lanka results in the accumulation of sodium salts in the soil layers and this factor is largely responsible for the abundance of the Na/K type in the dry zone. Further, the northern parts of Sri Lanka are underlain by sedimentary limestones, as a result of which the calcium type of water predominates in these parts. Increasing salinity has been observed in areas closer to the shorelines and in the Jaffna Peninsula in particular, this is commonly seen. The predominating anion in this type of water in the dry zone is Cl.

When one considers the topography, the central Highlands has groundwater of the Ca-$HCO_3$ type and with decreasing elevation, this merges into the non-dominant cation type. In the Lowlands the Na/K type predominates. Thus a Ca $\rightarrow$ NDC $\rightarrow$ Na/K type of sequence is apparent with decreasing elevations from the Highlands to Lowlands. This sequence could well be due to the different geochemical mobilities

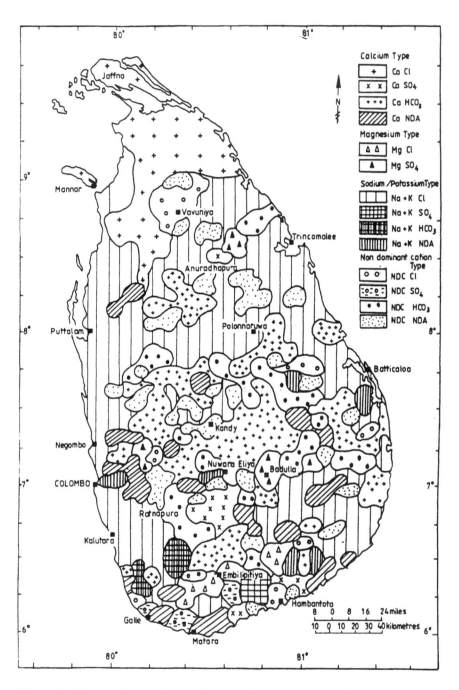

**Figure 2.** *The geochemical classification of groundwater in Sri Lanka.*

112

of the elements concerned. Further, there are numerous shallow and deep seated fractures and lineaments within the central regions of Sri Lanka and these are mainly responsible for the migration of groundwater within the hardrock terrains.

**The Geochemistry of Fluoride and the Incidence of Dental Diseases**

In a publication on rural water supply and sanitation in Sri Lanka, Peiris (1982) estimated that there are over one million wells on the island, of which 40% are used mainly for drinking and cooking purposes, 30% for bathing only and the balance for both. However, the percentage of safe water remains at 10-15% and greater emphasis is obviously needed in the development of safe water in Sri Lanka.

It is of interest to correlate the fluoride-rich and fluoride- poor areas delineated, with natural factors such as climate and geology (Figure 3). Low fluoride areas are situated mainly in the wet zone, whereas the high fluoride areas belong mainly to the dry zone. It is likely that in the wet zone where the average annual rainfall exceeds 500 cm, fluorides are easily leached. Fluoride is known to be easily leached from primary and secondary minerals and soils under the effect of high rainfall (Hawkes and Webb, 1962). In the dry zone regions, evaporation tends to bring the soluble ions upwards due to capillary action in soils. This, although not the sole explanation for the observed distribution of the fluoride in well water in Sri Lanka, could nevertheless be an important factor. However, it is the geology of the areas that needs special consideration. The composition of the rocks in the area, particularly the easily leached constituents, coupled with the climate, are the key factors in the geochemical distribution of elements in a tropical region. The abundance of fluoride in the rocks and the ease with which it is leached under the effect of groundwater has an important bearing on the abundance of fluoride in the areas concerned and hence the prevalence of dental diseases.

It is worthy of note that the high-fluoride zone of Sri Lanka lies on a mineralised belt at the Highland-eastern Vijayan geological boundary. Munasinghe and Dissanayake (1982) in their recent plate tectonic model for the geologic evolution of Sri Lanka suggested that the Highland-eastern Vijayan boundary is a mineralised belt. Fluorine being a volatile element is known to be abundant in such tectonic zones and is enriched in rocks found at such locations. Granites are generally rich in fluorine and such granites are found in abundance at the eastern Vijayan complex. Dissanayake and Weerasooriya (1986) studied the fluorine hydrogeochemistry on this tectonic boundary and confirmed the earlier findings that it is indeed a mineralised belt with abundant fluorides.

Recently with the drilling of nearly 7,000 deep wells mainly in the dry zone of Sri Lanka, more data on the fluoride distribution in the deeper circulating waters, has been obtained. It has been found that some of the deep wells in the North Central Province and in the areas around Uda Walawe had fluoride concentrations reaching levels as high as 10 ppm, well in access of the WHO recommended levels. Table 1 illustrates the fact that, in areas where the fluoride content is below normal, dental caries among the inhabitants are prevalent, whereas dental fluorosis is more common in the areas with a higher fluoride content (Dissanayake, 1979). Among the areas

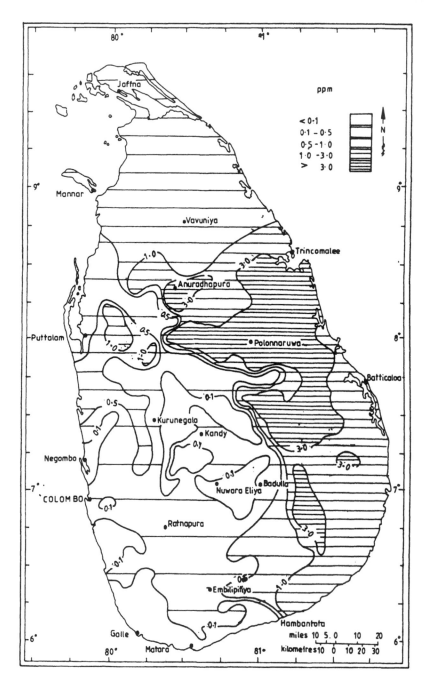

**Figure 3.** *Distribution of fluoride ions in the groundwater of Sri Lanka.*

114

**Table 1.** *The incidence of dental diseases in three cities of Sri Lanka.*

|  | Anuradhapura | Polonnaruwa | Kandy |
|---|---|---|---|
| Dental fluorosis | 77% | 56% | 13% |
| Dental caries | 26% | 27% | 96% |
| Fluoride concentration | 0.34-3.75 ppm | 0.26-4.55 ppm | <0.2 ppm |

Note: The maximum fluoride concentration in the Anuradhapura area was 9.0 ppm, 5.8 ppm in the Maha Oya and 4.8 ppm in Uda Walawe.

containing the highest fluoride concentrations in well water, the regions around Eppawela and Anuradhapura are prominent. Senewiratne and Senewiratne (1975) reported fluoride concentrations as high as 9 ppm in these regions, even in the dug wells. The abundance of fluoride which caused severe dental fluorosis among people of this area can be attributed to an abundance of fluorine in the rocks. It is significant that in this area occurs an economically exploitable deposit of apatite (fluoro-hydroxy phosphate) known to contain reserves of 23 million tons. Analysis shows the apatite to contain a fluorine concentration of 1.5 - 2.4% (Jayawardena, 1976). The areas around Maha Oya, Moneragala, Sevanagala and Uda Walawe, Hambantota, *etc.* also show high fluoride concentrations and particular attention should be given to the dental health of the people in these regions.

In a recent study, Tennakone and Wickremanayake (1987) showed that trace quantities of the fluoride ion can catalyse the dissolution of metallic aluminium in very slightly acidic or alkaline aqueous media. Potentially hazardous levels of aluminium could get leached from cooking utensils if fluoridated water or fluoride-rich food stuffs are used. This finding is very significant in view of the fact that aluminium is often correlated with the incidence of Alzheimer's disease and other nervous diseases.

The fluoride hydrogeochemistry of fluoride-rich areas is therefore of great importance in the epidemiology of Alzheimer's disease in Sri Lanka.

## Medical Geochemistry of Nitrates and Human Cancer in Sri Lanka

*Distribution of nitrates in the groundwater of Sri Lanka*

Figure 4 illustrates the distribution of nitrates in the groundwater of Sri Lanka and the maximum nitrate levels observed in the districts of Sri Lanka. In general, the average nitrate levels are below the danger level of 50 mg/l specified by the World Health Organisation. In Jaffna, however, these levels are exceeded (Dissanayake *et al.*, 1984). The nitrate concentrations appear to show a marked increase in areas of high population density, extensive fertiliser usage and in areas of the wet zone where atmospheric electric discharges are frequent. The type of farming practice and

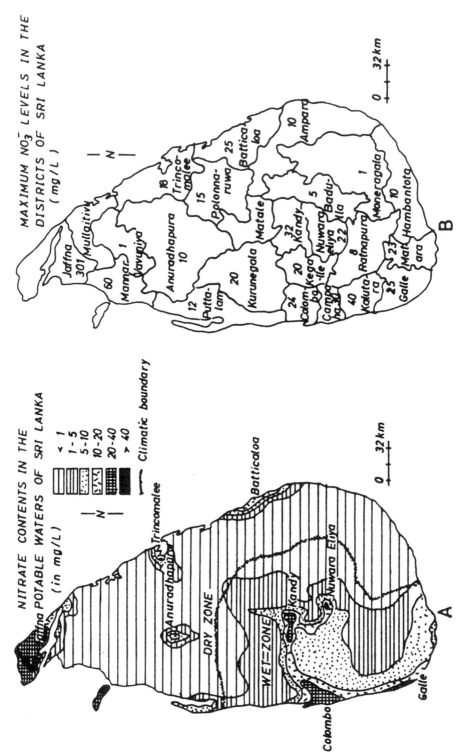

Figure 4. Distribution of nitrate ions in the groundwater of Sri Lanka.

116

fertiliser use is also an important factor in any research study concerning the regional distribution of nitrates.

In Sri Lanka, highly nitrogenous fertilisers such as urea are used in abundance, and farming practice and agriculture influence the abundance of nitrates in the water. As shown in Figure 4, the Jaffna peninsula has the highest nitrate contents in the groundwater of Sri Lanka. Geologically, the Jaffna Peninsula is underlain by highly fractured and karst limestone. There is a thin soil mantle of the red-yellow latosol type, and in the southern part of the peninsula are 10-20 m of fine sand which lie over the limestone formation. According to Gunasekaram (1983), 80% of the groundwater of the Jaffna peninsula is being extracted from the limestone aquifer and utilised for drinking, domestic, agricultural and industrial purposes, the rest being obtained from the sand aquifer. The water table in the Jaffna peninsula is very shallow on account of the surface aquifers.

The fact that Jaffna has nitrate levels exceeding the WHO limits by 100-150% is thus mainly due to the abundant nitrogenous waste matter in the form of human excreta and synthetic and animal fertilisers reaching the shallow groundwater table aided by the surface limestone aquifer. The geological conditions are therefore ideal for the excessive accumulation of nitrates. The poor sanitary conditions are mainly caused by improper planning of soakage pits and latrines and this aids in the serious contamination of the groundwater by nitrates.


## Incidence of Human Cancer in Sri Lanka

Panabokke (1984) in a five year study on the geographical pathology of malignant tumours in Sri Lanka presented data on investigations on 24,029 biopsy specimens. Accordingly, the Northern Province showed the highest incidence (184 per 100,000 population) of malignant tumours in biopsy material among the nine provinces of Sri Lanka. In the Southern Province, the incidence was low (37 per 100,000 population). The commonest sites from which malignant tumours arose were oesophagus (13.9 per 100,000 population), buccal region (12.2) and breast (10,58).

Table 2 shows the incidence of the different types of cancer in the nine provinces of Sri Lanka. The highest incidence of oesophageal, stomach, liver and small intestine cancer was observed in the Northern Province of Sri Lanka, whereas the incidence of bladder and kidney cancer appeared to be low in the same province. The incidence of oesophageal cancer, however, is high in Sri Lanka and is worthy of serious note. An interesting feature of the epidemiology of oesophageal cancer is the accumulation of substantial evidence indicating a relationship between geology, soils and climate with the incidence of oesophageal cancer, particularly in circumscribed areas (Laker et al., 1980; Kibblewhite, 1982; Laker, 1979).

Epidemiologists have established peculiar uneven distribution patterns for the incidence of oesophageal cancer in less developed rural areas where the average incidence of the disease is high (Laker et al., 1980), e.g. Transkei and the Caspian area of Iran. Laker et al. (1980) established an integrated model, the main theme of which was the relationship of the cancer incidence rate of the level to some mineral element in humans or in the crops which form their staple diet. An abnormal level of

**Table 2.** *Incidence of human cancer in Sri Lanka.*

| Province | Benign tumours | Malignant tumours | Bladder | Kidney | Small intestine | Stomach | Oesophagus | Liver |
|---|---|---|---|---|---|---|---|---|
| Northern | 92 | 184 | 0.90 | 0.30 | 1.30 | 5.80 | 37.4 | 3.3 |
| Northwestern | 61 | 63 | 0.40 | 0.40 | 0.30 | 1.20 | 4.6 | 0.78 |
| North Central | 35 | 24 | 0.20 | 0.20 | 0.00 | 0.90 | 1.26 | 0.36 |
| Eastern | 24 | 39 | 0.00 | 0.30 | 0.70 | 0.50 | 1.20 | 1.6 |
| Central | 114 | 84 | 1.00 | 0.50 | 0.10 | 2.20 | 13.5 | 1.9 |
| Western | 184 | 156 | 5.80 | 1.70 | 0.70 | 4.50 | 24.1 | 2.6 |
| Sabaragamuwa | 40 | 57 | 0.07 | 0.10 | 0.30 | 1.60 | 12.8 | 0.6 |
| Uva | 46 | 58 | 2.40 | 0.50 | 0.60 | 3.00 | 10.6 | 1.8 |
| Southern | 52 | 37 | 1.20 | 0.80 | 0.12 | 0.50 | 1.30 | 0.7 |

**Table 3.** *Average values NO⁻₃, NO⁻₂, Cl⁻ and TDS in the nine provinces of Sri Lanka.*

| Province | $NO_3^-$ (ppm) | $NO_2^-$ (ppm) | $Cl^-$ (ppm) | $TDS^*$ (ppm) |
|----------|---------|---------|--------|--------|
| Northern | 12.6 | 37 | 1263 | 1991 |
| Northwestern | 5.11 | 76 | 481 | 637 |
| North Central | 2.54 | 41 | 641 | 856 |
| Eastern | 4.98 | 94 | 740 | 816 |
| Central | 5.61 | 148 | 42 | 195 |
| Western | 9.02 | 189 | 58 | 201 |
| Sabaragamuwa | 4.15 | 212 | 12 | 128 |
| Uva | 1.75 | 210 | 47 | 346 |
| Southern | 5.25 | 49 | 321 | 410 |

*TDS: Total Dissolved Solids

such an element (deficiency or toxicity) is expected to cause physiological abnormalities in the human or the staple foodcrop, which lead to the production of carcinogenic substances. The mineral element level in the staple food depends on the level of availability of that element in the soil which, in turn, is dependent upon various soil factors, especially pH, which is itself determined by environmental factors.

**Nitrates and Cancer**

Table 3 shows the average chemical analysis of well water samples of the nine provinces of Sri Lanka. As illustrated in Figure 5, significant correlations are observed for certain types of cancer with the average nitrate concentrations. It is worthy of note that the nitrite concentrations of the water did not show any significant relationship with cancer incidence indicating the predominance of an *in vivo* reaction involving nitrite as against an *in vitro* reaction.

It is necessary at this juncture to emphasise the fact that even though there appears to be a significant correlation between the incidence of certain human cancers in Sri Lanka with the overall nitrate content in the groundwater, causative effects need not be attributed to the nitrate contents. Tannenbaum and Corea (1985) commenting on a paper by Forman *et al.* (1985) on nitrates, nitrites and gastric cancer in Great Britain noted that in the case of gastric cancer, the prevailing aetiologic hypothesis calls for a complex interaction of irritants, nutritional deficiencies, mucosal atrophy, bacterial overgrowth, *in vivo* nitrosification and sequential mutations which is so intricate that it represents a special challenge to the epidemiologist. A given dose of nitrate may be harmless to a normal subject but

119

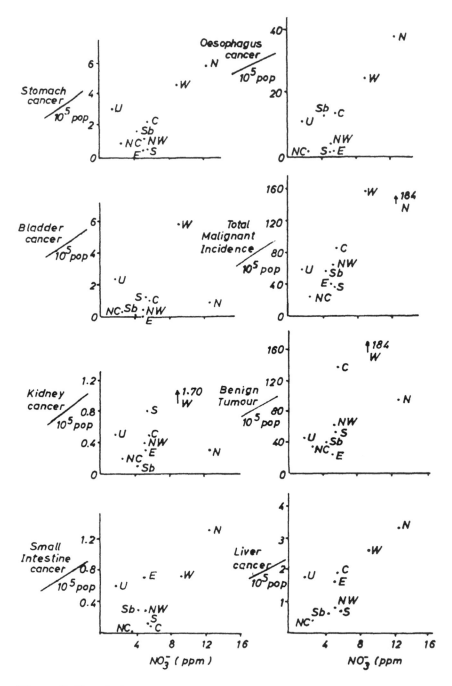

**Figure 5.** *Correlation of nitrate contents of groundwater with the incidence of human cancer in Sri Lanka. N = Northern Province; W = Western Province; C = Central Province; E = Eastern Province; U = Uva Province; NW = Northwestern Province; S = Southern Province; Sb = Sabaragamuwa Province; NC = North Central Province.*

noxious to a patient with atrophic gastritis, especially if the diet contains precursors of N-nitroso mutagens and carcinogens. The study of the geographical distribution of cancer in relation to nitrate concentrations in the geochemical environment must therefore necessarily be viewed in the light of the limitations of epidemiological research in understanding cancer. Forman *et al.* (1985) make the point that if nitrate exposure is a crucial factor in the development of cancer, epidemiologists can legitimately ask the question whether populations that experience cancer have a nitrate exposure, or alternatively whether populations exposed to a lot of nitrate, experience cancer.

## Endemic Goitre in Sri Lanka

The problem of endemic goitre in Sri Lanka has been investigated by many workers, the most recent study being that of Fernando *et al.* (1987) which shows that:
1. Endemic goitre in Sri Lanka is of relatively recent origin.
2. Endemic goitre is mainly due to iodine deficiency, while other factors such as genetics, geological characteristics, goitrogens, polluted water and trace elements may be contributory factors.
3. Western, Southern, Central, Sabaragamuwa and part of Uva Provinces of Sri Lanka, compose the endemic goitre belt with an estimated 10 million persons at risk.

The work of Fernando *et al.* (1987) also shows that the present endemic goitre belt should be enlarged to include the whole of Uva Province, and possibly the whole or part of the Northwestern Province. The available evidence shows that the etiology of endemic goitre is multifactorial while the chief causative factor is iodine deficiency in the diet. Work now being carried out by the authors appear to indicate that the geochemistry of Co and Mn also plays an important role in the etiology of endemic goitre.

## Acknowledgements

The authors wish to thank Prof. Cyril Ponnamperuma, Director, Institute of Fundamental Studies for his keen interest and useful discussions.

## References

Cooray, P.G. (1978). Geology of Sri Lanka. In: P. Nutalaya (ed.), *Regional Conf. Geology and Mineral Resources of South East Asia, Bangkok*, pp.701-710.
Dissanayake, C.B. (1979). *The Science of the Total Environment*, 13, 47-53.
Dissanayake, C.B. (1984). Environmental geochemistry and its impact on humans. In: C.H. Fernando (ed.), *The Ecology and Biogeography of Sri Lanka*, pp.65-97. Dr W. Junk Publishers, The Netherlands.
Dissanayake, C.B. and Ariyaratne, U.G.M. (1980). *Int. J. Env. Studies*, 15, 133-143.
Dissanayake, C.B. and Hapugaskumbura, A.K. (1980). *Indian J. Earth Sci.*, 1, 94-99.

Dissanayake, C.B. and Jayatilaka, G.M. (1980). *Water Air and Soil Pollution*, **13**, 275-286.

Dissanayake, C.B. and Weerasooriya, S.V.R. (1985). *The Hydrogeochemical Atlas of Sri Lanka*, p.103. Publication of the Natural Resources, Energy and Science Authority of Sri Lanka.

Dissanayake, C.B., Weerasooriya, S.V.R. and Senaratne, A. (1984). *Aqua*, **1**, 43-50.

Dissanayake, C.B. and Weerasooriya, S.V.R. (1986). *Chem. Geol.*, **56**, 257-290.

Fernando, M.A., Balasuriya, Herath, K.B. and Katugampola, S. (1987). Endemic goitre in Sri Lanka. In: C.B. Dissanayake and A.A. Leslie Gunatilaka (eds.), *Some Aspects of the Chemistry of the Environment of Sri Lanka*, (in press).

Forman, D., Al-Dabbagh, S. and Doll, R. (1985). *Nature*, **313**, 620- 625.

Gunasekaram, T. (1983). Groundwater contamination and case studies in Jaffna Peninsula of Sri Lanka. Paper read at the IGS- WRB Workshop, Colombo, Sri Lanka.

Jayawardena, D.E. de S. (1976). The Eppawela Carbonatite Complex in Northwest Sri Lanka. *Econ, Geol. Bull.*, **3**, 41. Geological Survey Department of Sri Lanka.

Kibblewhite, M.G. (1982). The influence of trace element distribution and availability in soils on the occurrence of oesophageal cancer. In: M.G. Kibblewhite and M.C. Laker (eds.), *Distribution in Relation to Oesophageal Cancer in the Butterworth District, Transkei*. University of Fort Hane, South Africa.

Laker, M.C. (1979). Mineral element studies on soil and plant samples from low and high incidence districts. In: S.J. Rensburg (ed.), *Environmental Associations with Oesophageal Cancer in Transkei*. Tygerberg Medical Research Council, South Africa.

Laker, M.C., Hensley, M., de L. Beyers, C.P. and Van Rensburg, S.J. (1980). *S. African Cancer Bull.*, **24**, 69-70.

Munasinghe, T. and Dissanayake, C.B. (1982). *J. Geol. Soc. India*, **23**, 369-380.

Panabokke, R.G. (1984). *Ceylon Med. J.*, **29**, 209-224.

Peiris, N.D. (1982). *Aqua*, **4**, 9-11.

Senewiratne, B. and Senewiratne, K. (1975). *Ind. J. Med. Res.*, **63**, 302-311.

Tannenbaum, S.R. and Correa, P. (1985). *Nature*, **317**, 675-676.

Tennakone, K. and Wickramanayake, S. (1987). *Nature*, **325**, 202- 203.

Vitanage, P.W. (1970). A study of the geomorphology and morphotectonics of Ceylon. *Proc. 2nd Seminar on Geochemical Prospecting Methods and Techniques*, E 72, II F.2, pp.391-405. UN, New York.

Weerasooriya, S.V.R., Senaratne, A. and Dissanayake, C.B. (1982). *J. Env. Management*, **15**, 239-250.

# 12 Aspects of Agricultural and Environmental Trace Element Research in the Republic of Ireland

Garrett A. Fleming
*An Foras Taluntais (The Agricultural Institute),*
*Johnstown Castle Research Centre, Wexford, Republic of Ireland*

## Summary

*Some details are given concerning research work on four trace elements - cobalt, molybdenum, selenium and iodine. Emphasis is placed on the importance of both solid and glacial geological patterns in delineating areas of trace element deficiency or excess. The role of soil manganese in regulating the availability of cobalt is highlighted. The extent of areas geochemically enriched in selenium and molybdenum is shown and the possible implications of excess selenium and molybdenum for human health stressed. In the case of iodine the effect of soil texture in determining soil iodine levels is demonstrated.*

## Introduction

Trace element research in the Republic of Ireland has concentrated primarily on practical problems arising at farm level. Some of the elements associated with these problems also have environmental implications. As some 85% of the country is devoted to grassland most of the research has naturally been in this area. Much effort has been expended in survey work of one kind or another, together with field trials involving measurements of herbage uptake of applied trace elements and studies on the various interactions which occur. In practice, trace element problems in grassland are related to animal health rather than to pasture production.

Use has also been made of soil and stream sediment analyses in regional mapping. These studies are ongoing and serve to pinpoint areas where more intensive investigations may be necessary. Because trace elements are not normally supplied in fertilisers the soil is the major source of supply. Geochemical considerations therefore are fundamental to any investigation.

123

This paper outlines some of the work which has been carried out in relation to four trace elements important in human and animal nutrition *viz.* cobalt, molybdenum, selenium and iodine.

## Geology

The bedrocks of the Republic of Ireland consist almost entirely of Palaeozoic and older strata. This forms an interesting comparison with Northern Ireland where Mezozoic and Tertiary structures are quite common. In the south east of the Republic, Ordovician and Silurian sediments predominate; the south and south west is composed mainly of Devonian sandstone together with some Carboniferous shales and limestones. Namurian rocks cover much of the area north and south of the Shannon estuary and further north in Counties Galway and Mayo granites, schists, gneisses and quartzites predominate. Granite and Dalradian schists and quartzites occupy much of the north west. The centre of the country is mostly underlain by Carboniferous limestone (Figure 5) with some Devonian sandstone in places. South of Dublin and trending in a north east-south west direction the Leinster granite extends for a distance of some 100 km. The country was subjected to a number of glaciations, the most recent - the Midlandian - abating some 12,000 to 15,000 years ago. The country was thus overlain with glacial deposits of varying depths and

**Figure 1.** *Glacial deposits in Ireland (Charlesworth, 1963).*

**Table 1.** *Cobalt content of soils formed from different parent materials.*

| Parent material | Number of soils | Co (mg/kg) Range | Mean |
|---|---|---|---|
| Basic igneous | 7 | 6.3-17.0 | 12.8 |
| Mica schist | 5 | 10.4-14.2 | 12.6 |
| Shale | 56 | 1.6-18.4 | 8.2 |
| Limestone | 275 | 1.8-17.5 | 6.0 |
| Sandstone | 75 | 0.5-13.8 | 3.6 |
| Gneiss | 6 | 0.2-4.4 | 2.4 |
| Granite | 79 | 0.3-17.5 | 2.1 |
| Blown sand | 19 | 0.2-1.0 | 0.4 |

textures ranging from the very tenaceous tills of the drumlin belt to gravels and sands further south. A glacial geology map of the country is shown in Figure 1.

The extent of morainic and other glacial deposits may be noted and these are relevant in interpreting soil trace element distribution. The geological events have led to a relatively complex soil pattern and it is not surprising that in many cases relatively small and well scattered pockets exist where trace element levels may be anomalously high or low.

## Cobalt

Cobalt in the form of vitamin B12 is required by animals and humans. In plants Co does not appear to be required by non- legumes, but as it is essential for nitrogen fixation by *Rhizobium* it is indirectly required by legumes. In practical agriculture its importance lies in the ability of different soils to supply sufficient of the element for efficient nitrogen fixation by clovers on the one hand and for ruminant nutrition on the other. In Ireland soils with inherently low levels of cobalt are formed mainly from acid igneous rocks *e.g.* granites and rhyolites, from sedimentary rocks such as sandstones and conglomerates and from metamorphic rocks as typified by quartzites, schists and gneisses. Peat soils are also inherently low in cobalt. Levels in limestone soils are very variable depending on the purity of the parent rock. Contents in shale soils range from low to quite high, depending on whether the shales are arenaceous (sandy) or argillaceous (clayey). Cobalt levels in Irish soils formed from a variety of parent materials are shown in Table 1. Sampling depth was 10 cm. Cobalt was extracted from soils with conc. HCl, thus the figures are essentially total values.

*Cobalt availability - the manganese factor*
For many years it has been known that cobalt availability in soils is affected by such factors as pH and drainage status (Mitchell, 1965). In the past decade or two evidence has been accumulating concerning the effect of iron and manganese oxides.

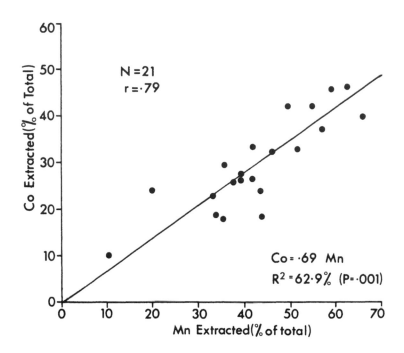

**Figure 2.** *Cobalt/manganese relationship in soil extracts.*

The effect of manganese oxides in particular has been highlighted by Australian researchers (McKenzie, 1967; 1970; Taylor, 1968) and in many instances manganese dioxide is seen as the primary sink for soil Co.

Manganese dioxide minerals often occur in soils in extremely finely divided forms and this results in their absorptive capacities being out of all proportion to their content by mass. The crystal structures of many of these minerals is such that $Co^{2+}$ ions are readily accommodated within the lattice. In short, the availability of cobalt in high Mn soils is severely restricted even in soils with high total Co contents.

The association between cobalt and manganese and cobalt and iron has been studied in Irish soils and findings point to a closer association between Co and Mn than between Co and Fe (Fleming, 1983). In a further study of the effect of Mn, soils formed from a variety of parent materials were extracted by the author with 10% hydrogen peroxide in 0.001 M nitric acid - a reagent which solubilises manganese dioxide in soils and used previously for similar studies in Australia (Taylor & McKenzie, 1966). The close association between Co and Mn in the extracts is apparent (Figure 2).

The findings are similar to those of Australian workers, and this is especially interesting as the genesis of the Australian and Irish soils was quite different.

The author has also investigated the effect of applied cobalt on the cobalt content of pasture on soils of differing total Mn status. Results are shown in Table 2.

**Table 2.** *Availability of cobalt from soils of differing manganese contents.*

| Soil Parent material | Limestone/ shale | Limestone/ shale | Granite | Limestone | Sandstone |
|---|---|---|---|---|---|
| pH | 6.5 | 5.6 | 6.6 | 6.4 | 5.8 |
| Org C (%) | 3.1 | 2.9 | 7.7 | 1.9 | 8.2 |
| Clay (%) | 24.0 | 22.0 | 11.0 | 8.0 | 9.5 |
| Fe (%) | 2.5 | 2.9 | 0.96 | 0.57 | 1.4 |
| Co (ppm) | 7.0 | 7.7 | 2.8 | 2.5 | 1.0 |
| Mn (ppm) | 940.0 | 540.0 | 160.0 | 55.0 | 25.0 |

| $CoSO_4.7H_2O$ (kg/ha) | Co (mg/kg) in grass | | | | |
|---|---|---|---|---|---|
| 0 | 0.02 | 0.04 | 0.03 | 0.05 | 0.11 |
| 1.32 | 0.04 | 0.07 | 0.11 | 0.17 | 0.82[*] |

[*] 1 kg/ha ($CoSO_4.7H_2O$) applied in this case.

It is apparent that the uptake of cobalt by grass is inversely proportional to the soil Mn content. Table 2 also shows that soil Fe generally parallels soil Mn but in view of the Australian studies it would seem that the prime scavenger for cobalt in soils is manganese dioxide in one of its many forms. The association of cobalt with iron oxides may be an apparent one as iron and manganese oxides commonly coexist in soils. They appear however to occur as discrete entities as electron microprobe studies on Canadian soils have shown (Brewer *et al.*, 1973). The practical implication of the "manganese factor" is that the benefit of applying cobalt to pastures in order to supply sufficient for grazing ruminants is contingent on the content of manganese in the soil.

### Molybdenum

Molybdenum is an element essential for both plants and animals. In Ireland shortage of Mo rarely occurs though instances of crop responses to applied Mo have been recorded on acid sphagnum peats in the midlands. By contrast many cut-away midland peats support molybdenum-toxic pastures. The high pH conditions resulting from underlying calcareous parent materials ensure that Mo is very freely available. Many mineral soils are also molybdeniferous and this can be traced to the nature of

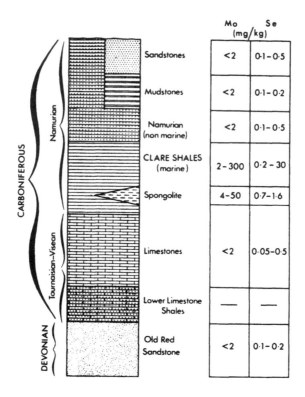

**Figure 3.** *Molybdenum and selenium contents of different rock strata.*

the parent material and to the drainage status of the soil. Molybdenum is much more mobile when drainage is poor. In such circumstances soil pores which would normally be filled with air become filled with water, with the result that oxygen supply becomes restricted. The reducing conditions then obtained favour the breakdown of molybdenum - bearing minerals and lead to increased availability of Mo. This increased availability leads to enhanced Mo levels in pastures on these soils and this in turn may give rise to problems in grazing animals.

With regard to parent material, marine black shales of Namurian (mid-Carboniferous) age would seem to be the main source of high Mo though Calp (earthy) limestones also contribute significantly. Figure 3 shows the concentration of molybdenum (and selenium) in strata of various ages from Co. Limerick (Atkinson, 1967).

The high contents of Mo and Se in the Namurian Clare shales are readily apparent. Namurian deposits in Ireland occur in a downwarp in the Carboniferous limestone extending in a Caledonoid direction from the estuary of the river Shannon to north of Dublin (Hodson and Le Warne, 1961; Charlesworth, 1963). Only the marine shale facies appear to be enriched in Mo and Se.

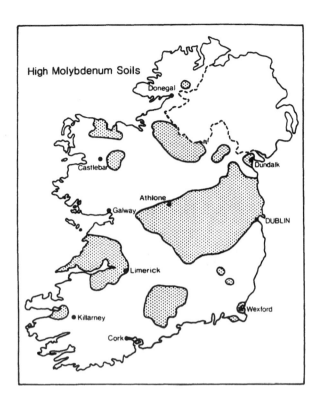

**Figure 4.** *Regional distribution of molybdeniferous soils in Ireland.*

*Regional distribution of molybdeniferous soils*
The regional distribution of molybdeniferous soils is shown in Figure 4. The figure has been compiled from the analysis of some three thousand soil, stream sediment and herbage samples and represents the position as presently known. Soils analysing greater than 0.3 mg/kg Mo extractable by 0.275 M ammonium acetate (Grigg, 1953) are regarded as molybdeniferous and as such are capable of supporting pastures with undesirably high levels of Mo for grazing animals.

Because of other variables, *e.g.* sulphur content of diet and level of soil ingestion, the Mo contents of pasture liable to cause problems in animals are difficult to quantify but values in excess of 5 mg/kg Mo give cause for concern. Various factors influence the Mo content of pastures including species and season. Clovers may contain up to five times as much Mo as ryegrasses in the same sward and the latter frequently contain more Mo in the autumn and early winter, than spring or summer. This is possibly due to a natural increase in soil moisture which occurs at this time.

Although Figure 5 represents solid geology and as such takes no account of glacial dispersion patterns, there is an obvious relationship between the occurrence of muddy limestones and molybdeniferous soils. Some relationship also exists between the occurrence of Namurian rocks and molybdeniferous soils but this relationship

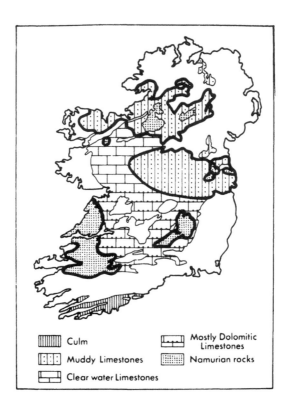

 shows a map with the following legend:

|  | |  | |
|---|---|---|---|
| ▥ | Culm | ⊞ | Mostly Dolomitic Limestones |
| ⊡ | Muddy Limestones | ▦ | Namurian rocks |
| ⊟ | Clear water Limestones | | |

**Figure 5**. *Distribution of Carboniferous rocks in Ireland.*

can only be tenuous as, in the Namurian deposits, only the marine black shales appear to be enriched in molybdenum (Figure 3).

*Significance of molybdeniferous soils*
From the agricultural point of view the incidence of molybdeniferous soils and pastures is of importance in relation to copper deficiency in animals. When molybdeniferous forage is ingested by grazing animals the molybdenum acts as an antagonist preventing optimum absorption and storage of copper. This molybdenum-induced copper deficiency is especially severe in young animals. In the case of horses, copper shortage is associated with leg bone malformation, and, as many of the most extensive stud farms in Ireland are located in the high molybdenum belt, the mapping of molybdeniferous soils is of obvious importance in relation to the bloodstock industry. The swayback syndrome in lambs stems from a copper deficiency in the pregnant ewe who must receive timely supplementation with copper in order to prevent the condition occurring in the newborn. From the human health aspect, high molybdenum in drinking water may have some effect on the incidence of dental caries.

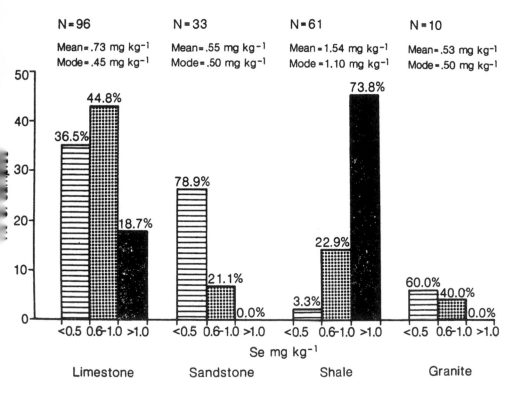

**Figure 6.** *Frequency distribution of selenium in soils formed from different parent materials.*

### Selenium

The geochemistry of selenium is in many respects similar to that of molybdenum. Marine black shales form the main parent material of seleniferous soils and the elevated levels which occur in these rocks are evident from Figure 3. It is very likely that selenium occurs as a guest element in ferrous sulphide crystals in the shales but some is probably organically bound.

In Irish soils selenium differs from molybdenum in being far less dispersed. Although extremely high levels occur in some soils these are found in small pockets and are typically low lying, poorly drained, organic matter rich and neutral to alkaline in reaction. Selenium, leached from black shales and Calp limestones and from glacial overburden enriched with these rocks, reaches depressions filled by river flood plains and old lake beds and it is here that the most selenium-toxic soils occur. Total selenium in these soils ranges from 5 to over 1,000 mg/kg (Fleming, 1962) but the majority of seleniferous soils contain between 5 and 50 mg/kg Se.

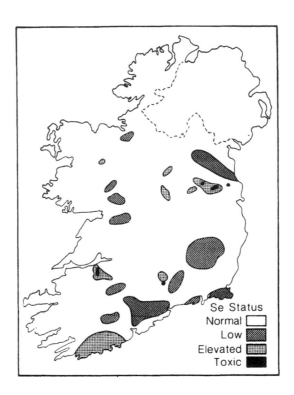

**Figure 7.** *Selenium in Irish soils.*

*Selenium in different soils*
The frequency distribution of Se in soils formed from four different parent materials
- granite, shale, limestone and sandstone is shown in Figure 6.

The soils formed from sandstone and granite contained no selenium in excess of
1 mg/kg but the number of granite soils examined was small. Normally acid igneous
rocks and the soils formed therefrom are quite low in Se content. Indeed many
selenium-deficient soils originate from acid igneous parent materials. The soils
formed from limestone, and - particularly - shale, contained the highest levels of Se.
The following point must, however, be noted. The terms "limestone" and "shale" are
by their nature general. Many types of limestone occur ranging from very pure
calcium carbonate to dolomitic limestones and to those such as Calp which have
been contaminated with foreign material. With the exception of those trace elements
which can be easily accommodated into a calcium carbonate lattice the trace element
content of limestones is primarily a function of the degree of impurity of the rock.
Selenium is no exception here. Likewise, shales vary from arenaceous or sandy types
to argillaceous or clayey types and the latter are richer in trace elements. The data of
Figure 6 refer to soils taken on a countrywide basis and as such would have formed
from limestones, sandstones, shales and granites of varying composition.

**Table 3.** *Selenium and organic matter in a seleniferous soil profile.*

| Soil depth (cm) | Organic matter (%) | Se (mg/kg) |
|---|---|---|
| 0-15 | 31.0 | 20.0 |
| 15-30 | 75.0 | 75.0 |
| 30-50 | 6.6 | 6.4 |
| 50-60 | 62.6 | 100.0 |
| 60-85 | 8.0 | 2.7 |

*Geographical distribution of selenium in Irish soils*
The geographical distribution of selenium in Irish soils as presently known is shown in Figure 7.

Total Se values from surface soil samples (0-10 cm) were used in compiling the map. Values of less than 0.5 mg/kg Se are regarded as low, 0.6 to 1.4 mg/kg as normal, 1.5 to 5 mg/kg as elevated and those greater than 5 mg/kg Se as toxic.

The incidence of toxic soils is definitely related to known seleniferous parent materials and those soils in the low category are associated to some extent with morainic and general sandy deposits. However, the closeness of this association has not been well established to date and must await further survey work.

*Selenium and organic natter*
Mention has been made previously of the possible association between selenium and organic matter in black shales. In Irish soils this association is particularly obvious (Table 3). It must be stressed however that contents have been expressed on an air-dry basis and the bulk-density of the various horizons has not been taken into account.

The accumulation of organic matter in the 50-60 cm horizon of the profile may appear puzzling. The profile in question was taken from an area which, after initial soil formation, was subsequently reflooded - probably by glacial meltwaters - and a second cycle of soil development ensued.

*Some practical considerations*
As selenium is not required for plant growth, shortage of the element does not affect pasture production. Neither apparently does selenium excess. Animal health however is affected. In cases of selenium shortage animals may suffer from a variety of syndromes ranging from muscular dystrophy and white muscle disease in *e.g.* lambs to retained placentae in dairy cows. In Ireland the latter syndrome seems to be the only one clearly manifest; white muscle disease, for example, has not been reported.

Selenium excess has affected horses and cattle and the well known symptoms of loss of hair and cracking of hooves have been recorded (Fleming and Walsh, 1957). Selenium toxicity is not important on a national scale. It is confined to relatively small areas but for some individual farmers it constitutes a distinctly troublesome problem. In the human health sphere selenium may be a factor in the incidence of dental caries. Some epidemiological studies are currently being initiated involving school children from seleniferous and non-seleniferous areas.

## Iodine

Because of a high incidence of human goitre, investigations into the iodine content of some Irish soils were conducted in the 1930s and 1940s (Shee, 1940; O'Shea, 1946). Soils from a goitrous area averaged 3.1 mg/kg I whereas those from an area where goitre was virtually unknown averaged 40 mg/kg I. The latter area was on the western seaboard and seaweed was commonly used as a fertiliser at that time. Subsequent investigations by the Iodine Education Bureau, London in collaboration with local veterinary practitioners, animal nutritionists and agricultural advisors led to some delineation of iodine deficient areas (Figure 8).

The investigations involved the provision of iodised mineral supplements to stock on over three hundred farms and observation of resultant effects on animal health and performance.

As iodine is readily leached from soils it is natural to focus on coarse-textured, freely drained soils as sources of low iodine. Such soils are frequently associated

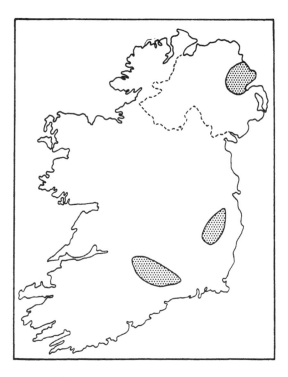

**Figure 8.** *Some areas of low soil iodine in Ireland.*

with glacial and fluvio- glacial deposits, *e.g.* moraines and outwash plains. It is interesting and probably not coincidental that the areas designated as low in iodine by Shee (1940) and O'Shea (1946) lie on the end moraine of the Midlandian glaciation (Figure 1). Examination of glacial deposit patterns could prove very useful in endeavouring to locate other low-iodine soils. In counties where a soil survey has been completed information will obviously be more precise and the task of identifying low iodine areas should be that much easier.

It is remarkable that until recently (McGrath and Fleming, unpublished) no further investigations of iodine in Irish soils have been undertaken. A partial explanation is the fact that human and animal goitre can be so easily prevented by supplying iodine; to humans in the form of iodised table salt, and to animals in the form of salt licks and mineral mixes.

*Soil iodine and distance from the sea*
It is generally recognised that atmospheric precipitation is a most important source of soil iodine (Fuge and Johnson, 1986). Because of its solubility, iodine is relatively enriched in both surface and sea waters and onshore winds carrying sea-spray contribute to the iodine content of soils near coasts. Iodine deficiency is classically associated with areas remote from the sea but, whereas this may hold for large continental masses, it is not necessarily so for relatively small islands such as Ireland. Fleming (1980) has drawn attention to the importance of soil type in any assessment of iodine in soils. Levels of iodine in heavy-textured inland soils can be higher than those in light- textured soils closer to the sea. A recent study has borne this out (McGrath & Fleming unpublished). Eighteen samples from a sandy soil near the sea were examined for total I content and compared with a similar number taken from a heavier-textured soil further inland. Results are shown in Table 4.

### Conclusion

The foregoing indicates some approaches and strategies adopted in trace element studies in Ireland. It has become increasingly evident that, in order to maximise the value of this and other trace element research, an interdisciplinary approach must be

**Table 4.** *Soil iodine and distance from the sea.*

| Soil type | Clay (%) | Mean distance from sea (km) | I (mg/kg) Range | Mean[*] |
|-----------|----------|------------------------------|------------------|---------|
| Sandy loam | 12.0 | 3.8 | 2.9-6.8 | 4.8 |
| Clay loam | 26.0 | 27.0 | 6.6-24.0 | 15.6 |

[*] Mean of 18 samples

135

adopted. The growing awareness of the importance of trace elements in health and disease underlines this.

The existence in Ireland of areas geochemically enriched in *e.g.* molybdenum and selenium, and already shown to be of importance in animal nutrition, suggests that these and possibly other trace elements may be exercising some effect on human health in these areas. Currently there are indications that selenium and possibly molybdenum are related to some dental problems in schoolchildren in at least one location. Clearly this must be investigated in more detail.

It is also evident that soils with low trace element contents are not uncommon in the country. From the animal health aspect the importance of these soils insofar as they support pastures with sub-optimal trace element levels has been highlighted. However in the tillage areas which are predominately in the south and east of the country the levels of - for instance - selenium and zinc in cereal grain may have relevance in both human and animal health. Recently zinc deficiency in cereals has been identified on soils mainly formed from fluvio-glacial deposits.

Information on trace element levels in Irish diets in general is not plentiful and work in this area is urgently required. Here emphasis should perhaps be concentrated on selenium and zinc, the former in view of its antioxidant role in cancer and immune system research; the latter because of its reported association with the development of foetal abnormalities. The establishment of reliable data bases for trace elements in diets is essential in any form of interdisciplinary approach to problem solving in the field of human nutrition.

## Acknowledgements

Thanks are due to Ms E. Spillane for preparing the text, to Mr P.J. Parle for proof-reading, and to Mr J. Lynch for the figures.

## References

Atkinson, W.J. (1967). Regional Geochemical Studies in County Limerick, Ireland with special reference to Selenium and Molybdenum. Unpub. Ph.D. Thesis pp.337.
Brewer, R., Protz, R. and McKeague, J.A. (1973). *Can J. Soil Sci.*, **53**, 349-361.
Charlesworth, J.K. (1963). In: *Historical Geology of Ireland*, p.283. Oliver O'Boyd, London.
Fleming, G.A. (1962). *Soil Sci.*, **94**, 28-35.
Fleming, G.A. (1980). In: B.E. Davies (ed.), *Applied Soil Trace Elements*, pp.205-206. John Wiley and Sons, New York.
Fleming, G.A. (1983). In: S.S. Augustithis (ed.), *The Significance of Trace Elements in Solving Petrogenetic Problems and Controversies*, pp.731-743. Athens Theophrastus Publications, S.A.
Fleming, G.A. and Walsh, T. (1957). *Proc. R. Ir. Acad.*, **58B**, 51- 166.
Fuge, R. and Johnson, C.C. (1986). *Environ. Geochem. and Health.*, **8**, 31-54.
Grigg, J.L. (1953). *N.Z. J. Sci. Technol.*, **34A**, 405-414.
Hodson, F. and Le Warne, G.C. (1961). *Quart. J. Geol. Soc.*, **117**, 307-333.

McKenzie, R.M. (1967). *Aust. J. Soil Res.*, **5**, 235-246.

McKenzie, R.M. (1970). *Aust. J. Soil Res.*, **8**, 97-106.

Mitchell, R.L. (1965). In: F.E. Bear (ed.), *Chemistry of the Soil.*, pp.320-368. Reinhold, New York.

O'Shea, E.M. (1946). *Ir. J. med. Sci. 6th Series* No.251, 748-752.

Shee, J.C. (1940). *Scient. Proc. R. Dubl. Soc.*, **22**, 307-314.

Taylor, R.M. (1968). *J. Soil Sci.*, **19**, 77-80.

Taylor, R.M. and McKenzie, R.M. (1966). *Aust. J. Soil Res.*, **4**, 29-39.

# 13 Effects of Some Trace Element Deficiencies and Toxicities on Animal Health in India

S.P. Arora
*Division of Dairy Cattle Nutrition, National Dairy Research Institute,*
*Karnal-132001, India*

## Summary

*Geochemical and soil mapping studies indicate that Zn and Cu deficiencies and Se excess are prevalent in certain zones of India. Feeds and fodders are therefore deficient in Zn and Cu and fodders like rice straw contain toxic levels of Se leading to Degnala disease. Requirement of dietary Zn for different species of livestock is 80 ppm. Sulphate treatment (Degcure) has been evolved to treat diseased cases of Se toxicity.*

## Introduction

Ailments as a result of micro-element deficiencies or toxicities in livestock and poultry are prevalent in many developing countries with showing of both frank and subdued clinical syndromes. Packages of practices developed for high yielding varieties of some crops like wheat and rice under intensive cultivation systems have led to depletion of certain micro- elements from certain geographical zones. This has resulted in a lesser dietary supply of certain trace elements than the requirement to livestock and poultry. The problem seems to be more of a geochemical nature leading to metabolic disturbances resulting in certain pathological changes in tissues and even manifestation of actual disease symptoms. Their biological interactions, because of physico-chemical resemblances, are so complex that an excess of one micro-element may suppress the absorption or utilisation of some other element antagonistically in tissues. In this review paper certain micro-elements have been selected for presentation because of reports of their deficiencies or toxicities from India and even from certain other developing countries during the recent past.

# Zinc

In certain geographical areas where intensive cultivation is being practised to get two or more crops/year under irrigated or otherwise conditions, the land has shown Zn as one of the micro- elements highly depleted from the soils (Kanwar and Randhawa, 1967). More than half of the soil samples collected from different regions of India under the ICAR co-ordinated scheme on micro-nutrients in soil and plant showed Zn deficiency ranging from 0.03 to 10.4 ppm. Zinc deficiency is greater in calcareous soils due to an increase in pH as a result of a greater amount of calcium salt. The availability of Zn is greatest at the soil surface and it decreases as a function of soil depth (Lal *et al.*, 1960). Soils having a pH ranging from 5.5 to 7.0 will provide more Zn to the plants. However, 1 ppm or less Zn in the soils can be considered as a critical level for most of the crops. The role of Zn is not only important from the crop production point of view, but also it is equally essential in animal nutrition. In spite of soil Zn fertilisation being practised in some areas, its content in feeds and fodders does not exceed 30 ppm. This dietary Zn level is too low to meet the requirements of livestock for production traits (Ott *et al.*, 1965). In body tissues Zn is in highest concentrations in skin, hair and wool. Its normal level in the blood serum is 800 µg/100 mL, though it has been reported as varying from 250 µg upwards (Underwood, 1977). In the internal organs Zn is at maximum concentrations in the prostate and pancreas. It seems that there is some control of Zn absorption from the intestine which is probably through metallothionein because, even with very high dietary doses of Zn, its mucosal uptake is limited (Bremner, 1981). The half-life of metallothionein varies from 1.4 days in liver to 4.2 days in kidney and its synthesis depends upon L-methionine availability (Whanger *et al.*, 1981). Zinc activates a number of enzymes in the body tissues involved in different biochemical transformations. The earliest report of this metal activating carbonic anhydrase shows it to be responsible for the transfer of carbon dioxide from tissues to the exhaled air. In carbonic anhydrase the Zn content is 0.3%. Other important enzymes the which are activated by Zn are glutamic dehydrogenase, alcohol dehydrogenase, alkaline phosphatase, carboxipeptidase, superoxide dismutase, lactic dehydrogenase *etc*. Another important function of Zn is in the utilisation of vitamin A through the activation of the Zn metalloenzyme alcohol dehydrogenase. As shown in recent evidence, β-carotene is also converted more into vitamin A in the intestinal wall through activation of alcohol hydrogenase (Prasad and Arora, 1978). Zinc also helps in the greater synthesis of retinol binding protein in the liver which later helps in the greater mobilisation of vitamin A from the liver hepatocytes into blood.

Zinc and vitamin A are a requirement in the transformation of light energy into electrical impulses in the retina later transmitted through optic nerves to the brain. Any disturbance in the process would lead to "Night Blindness", a disease which has been recorded either due to Zn deficiency or vitamin A deficiency (Prasad and Arora, 1979). It has been further observed in a number of species that the blood vitamin A level in serum will always remain low whenever there is dietary deficiency of Zn less than 30 ppm. For optimum mobilisation of vitamin A from the liver, Zn is required to the extent of 80 ppm as already reported in different species (Saraswat and Arora, 1972; Chhabra *et al.*, 1980). Just as "Night Blindness" can be induced by Zn

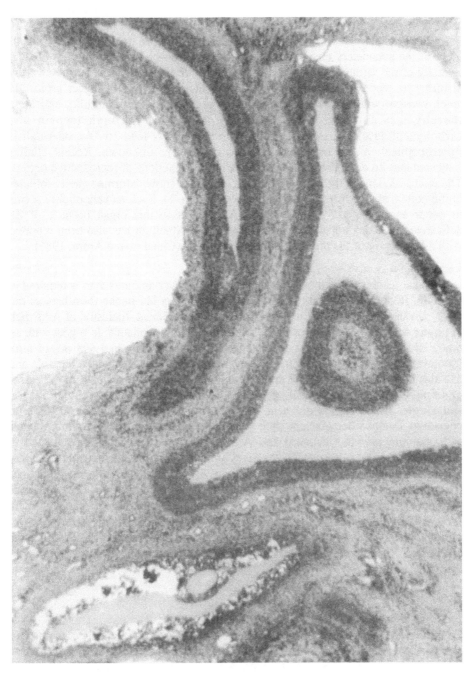

**Figure 1**. *Atretic follicle in the ovary of a goat.*

deficiency or vitamin A deficiency, similarly some animals show skin keratinisation which is probably due to disturbance in the synthesis of sulphated-mucopoly-saccharide in which both Zn and retinoic acid are a requirement (Gupta et al., 1972). From a diagnostic point of view Zn status in the animal can be diagnosed by one or more of the parameters such as serum Zn, alkaline phosphatase activity, Zn binding capacity of the serum and $^{65}$Zn in vitro uptake of erythrocytes. Another noteworthy disturbance, recorded due to its deficiency, is that rats exhibit a higher proinsulin level, whereas with normal Zn supplemented diets there is more insulin, indicating thereby less conversion of proinsulin to insulin through trypsin and carboxypeptidase. Impaired Zn absorption can also lead to "Acrodermatitis Enteropathica", a genetical fault, though details are unknown. Recent studies indicated that Zn deficiency can result in failure of functions of reproductive organs. The micro-sections of the ovary in Zn deficient goats show deformed atretic follicles in the cortex, along with some deformed ova (Figure 1). Such an abnormality seems to occur as a result of lower vitamin A supplies to this organ because of Zn deficiency. In males impaired testicular germinal epithelium has also been reported with a hampering of the process of spermatogenesis (Chhabra and Arora, 1985).

*Requirement*

In neonatal feeding Zn is thought to be secreted in lower amounts than is required in the milk. In cows Zn in milk is correlated with dietary Zn intake (Neathery et al., 1973; Underwood, 1977) but its level is low and insufficient. Inclusion of feeds rich in phytic acid are known to reduce absorption of Zn in neonatals. In a goat with an intake of 85 mg Zn/day, the quantity secreted is only 3.6 mg in 1,598 mL of milk (Sonawane and Arora, 1975). Zinc content in buffalo milk is also quite variable and has also been reported up to 5.7 ppm. In the colostrum, Zn is secreted to the extent of 48 mg/L with an intake of 50 ppm Zn. Zinc is associated with the casein fraction and citrate. If the diet is deficient in Zn, then its quantity is only 40 mg/L in colostrum. Lambs when given 80 ppm Zn show normal blood serum vitamin A levels and optimum growth (Saraswat and Arora, 1972). Further studies conducted in calves and goats revealed that Zn at a level around 80 ppm in the diet was necessary for maintaining normal blood serum vitamin A level and normal tissue functions (Arora, 1981; Chhabra and Arora, 1985). In human beings 16 mg Zn/day is the requirement for an adult. Women during lactation would require 20 mg Zn/day. In some calves of Friesian descent the Zn requirement has been estimated to the extent of more than 100 ppm in individuals suffering from a genetic anomaly. Severe cases of this genotype abnormality were treated with as high as up to 1,000 mg Zn/day for recovery (Mills et al., 1976).

# Copper

Copper is closely associated in its function with Fe. Copper deficiency is rampant in specific geographical zones. In cereals, straws and green fodders its range has been recorded to vary from 4.8 to 22.11 ppm. A level of 0.2 ppm available Cu has been fixed as the critical limit for adequacy of this mineral as a plant nutrient whereas the values in some of the soils in certain countries including India are less than 0.2 ppm and do not provide adequate quantity to plants and then in turn to feeds for livestock. Geochemically Mo is in excess in soils and leguminous crops take it up more, resulting in higher than recommended increases in the Cu requirement.

Copper, just like Fe, is also deficient in milk. Therefore, neonatals are more prone to anaemia as a result of Fe and Cu deficiency when they are kept on milk alone. Copper deficiency in experimental animals is liable to produce a number of heart abnormalities such as cardiac hypertrophy (Mills et al., 1976) and aortic aneurysms (O'Dell et al., 1961). In rats it is shown that catecholamine synthesis drops; this is the sum of noradrenaline, adrenaline and dopamine (Hesketh et al., 1985).

Copper is not a constituent of haemoglobin but it acts as a catalyst in the synthesis of haemoglobin. It is present in liver as hepatocuprein whereas in blood it is present as haemocuprein. Plasma ceruloplasmin activity and plasma Cu concentrations are widely used as indicators of Cu nutrition. In addition to haemoglobin synthesis, Cu is an activator of certain cellular enzymes which function in oxygen transport in the cells. Cu (II) can react with membrane sulphydryls and oxygen to form the superoxide ion ($O^-_2$) and $H_2O_2$ (Kumar et al., 1978). It also catalyses the oxidation of isolated oxyhaemoglobin to form the superoxide ion and methaemoglobin (Rifkind, 1974). Superoxide dismutase activity in erythrocytes varies from 0.2 to 0.5 mg SOD/g Hb and may indicate Cu status in chronic cases. Its activity drops rapidly during its deficiency in growing animals and it can be considered as an index of Cu status. SOD activity can drop to as low as 0.47 enzyme units/mg protein in Cu deficiency as compared to normal values of 1.90 enzyme units/mg protein at 20 ppm dietary Cu level (Sharma, 1982).

In ruminants, the blood Cu level ranges from 60-144 μg/100 mL (Sharma and Prasad, 1983). In its deficiency anaemia is usually of hypochromic and normocytic type. Blood Cu level cannot be used as a diagnostic tool of its deficiency because it is likely to remain normal in apparently healthy animals and may not show any health problem despite a low Cu status.

With Cu deficiency, cysteine is found more in rough wool than in fine quality wool. Copper deficiency disease, e.g. Swayback has been recorded in India in kids (Figure 2) which is indirectly due to excess Mo uptake by certain fodders (Prasad et al., 1982), symptoms very similar to a disease recorded in Australia in lambs. It takes two forms, one congenital and the other in which syndromes appear 6-8 weeks after birth. Usually degenerative changes occur in the collagen and elastin matrix of tendons leading to leg paralytic syndromes. In adult cattle in Australia a disease known as falling disease has been recorded due to Cu deficiency. Its deficiency has also been surveyed in Malaysia and the incidence is as high as 80% in some areas. Degenerative changes take place in the sympathetic nervous system of Cu deficient cattle. In young cattle there are changes in the texture and colour of the hair coat, black hair developing a grey bronze cast. Some may develop depigmentation of hair around eyes due to its deficiency. Effect of its deficiency upon skeletal structure is particularly reflected in the metatarsal bone, relatively narrower at mid shaft and wider at the epiphysis. Molybdenum forms Cu tetrathiomolybdate rendering Cu unabsorbable, and thus it is important to feed excess dietary sulphate for forming a

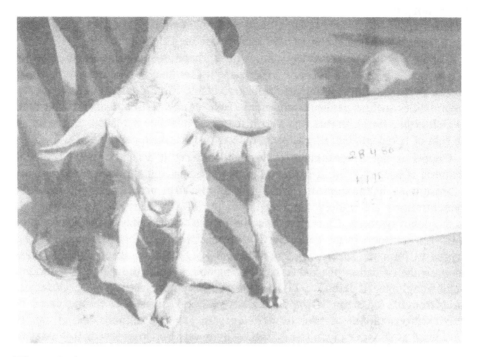

**Figure 2.** *A kid showing the syndromes of swayback.*

complex with Mo so that Cu is free to be absorbed. Alternatively, excess Cu is required to be supplemented.

In enzootic ataxia in lambs, as a result of Cu deficiency, certain histological changes of brain and spinal cord have been recorded as a result of cytochrome oxidase diminished activity and degradative changes in myelin strands (Smith *et al.*, 1981) or even showing patchy cerebellar cortical hypoplasia (Howell *et al.*, 1981). Another interesting phenomenon which has been reported is faulty Cu transport as a result of genetic error in tissues in which death occurs in children before three years of age, called "Menke's disease". Major pathological changes are the degeneration of arteries and brain damage without the syndromes of anaemia. Copper retention in the liver and overflow to brain in Wilson's disease of sheep has also been reported as a genetic problem. Cirrhosis of liver and neurological disorders appear in lambs due to non-availability of Cu to its metallo-enzymes (Danks, 1981).

*Requirement*
The copper requirement of cattle weighing 400 kg is 50 mg/day and in chicks its requirement is 2 mg/kg diet. The copper requirement of wool sheep is 10 mg/day. Kids suffering from Swayback can be treated by administering a 20 mL solution per day orally of 2 g copper sulphate dissolved in 1 L of water (Prasad *et al.*, 1982).

## Sulphur

Many crops in certain geographical areas of India under intensive cultivation are responding to S fertilisation giving higher yields (Merck and Dev, 1979). Its dietary importance in livestock is therefore being realised for adequate supplementation. Sulphur is present in the body tissues and wool up to the extent of 0.15% and it is utilised mostly as sulphate, cysteine, methionine and cystine. Methionine, apart from being utilised in protein synthesis, is a precursor of S-adenosyl L-methionine formed by the enzyme S-adenosyl methionine synthetase in liver. This compound is the methyl donor for a large number of macromolecular methyl transferases including DNA, RNA, proteins and phospholipids. Its importance in body metabolism through certain hormones such as glutathione and insulin is well recognised. Similarly, in the rumen microbes are able to synthesise thiamine and biotin which contain S in their structures. If S is available from the diet in sulphate form, the rumen microbes are also able to synthesise cystine and methionine, the latter being an essential amino acid. In the cartilage it is utilised as choindroitin sulphate.

Its deficiency leads to more lactic acid production in the rumen. Usually microbes require it in the ratio of 1:10 with nitrogen for best fermentation rate (Arora *et al.*, 1974). Adequate availability of S in certain body tissues is necessary for the synthesis of metallothionein, a protein having a large metal- binding capacity for transport of zinc, copper and other metals. Adequate S in the form of sulphate helps to antagonise the toxicity of molybdenum and, indirectly, thus frees copper for absorption to prevent copper deficiency. Both cystine and methionine can antagonise seleno-cystine and seleno-methionine at the cellular level to prevent selenium toxicity.

### *Requirement*

The sulphur requirement in the diet should be about 0.2% on a dry matter basis. Sulphur can be best supplemented through ammonium sulphate in the diet of ruminants which provides both N and S and helps the microbes in the rumen to synthesise more protein and increase cellulose digestibility (Ahuja and Arora, 1981). Sodium sulphate is an alternative source of S for dietary supplementation. Sulphur excretion as estimated in cattle and buffaloes varies from 1.5 to 2.0 g/day with an estimated endogenous component of 0.8 to 1.0 g/day against an intake of 3.7 to 4.5 g/day. The apparent digestibilities of S are about 55% in both the species, whereas true digestibility is about 80%. Using the multiple regression analysis, the maintenance plus growth requirements of S have been computed as 16.47 mg/kg body weight at peak level which corresponds to 0.13% in the diet. This is expected to give a growth rate of 450 g/day. Further estimates indicate that calves consume 3.03 kg dry matter/day with an average live weight of 102.67 kg. The requirements for different growth rates are as shown in Table 1.

The sulphur requirement for milk production has also been estimated and its ratio is 4.43 g S intake for 1 g S secreted in milk. Therefore, for milk production, the S requirement can be calculated by the following equation (Ahuja and Arora, 1980):

S requirement/day = 0.00175 (g) x live weight (kg)
+ 1.330 (g) x milk yield (kg).

145

**Table 1.** *Sulphur requirements for different growth rates.*

| Growth (g/day) | S requirement % DM |
|---|---|
| 400 | 0.13 |
| 500 | 0.15 |
| 600 | 0.16 |

## Selenium

Geochemical survey reports from different countries of the world reveal that Se is usually deficient in soils since it is a requirement in the activation of glutathione peroxidase in different body tissues. This enzyme carries four apparently identical sub-units of about 20,000 daltons each, and its molecular x-ray crystallography has demonstrated that one Se atom is bound to each sub-unit, present as selenocysteine (Kraus *et al.*, 1983). This enzyme is also used to assess the Se status of animals indicating risk of its deficiency. Its main role in tissues is to prevent the formation of peroxidase and oxidative damage from unsaturated fatty acids and thus maintaining the structure and functional integrity of biological membranes. From a therapeutic treatment point of view, Se and vitamin E have a mutual sparring action to a certain extent. Selenium deficiency can lead to a number of diseases which have been recorded in different species. "White Muscle Disease" in calves, "Stiff Limb Disease" in lamb, "Encephalomalacia" in chicks and muscular dystrophy in general have been named Se or vitamin E deficiency diseases. Mostly there are pathological changes of hyaline degeneration and coagulative necrosis in muscle cells leading to myopathy. The selenium requirement as a dietary essential is 0.1 ppm though certain reports indicate its requirements as high as 0.3 ppm in swine. Selenium is also a requirement for increased microbial protein synthesis in the rumen (Khirwar and Arora, 1976).

*Toxicity*
Whereas Se is a dietary essential reported by many groups of scientists from a number of countries, in some countries its soil level is much higher leading thereby to its accumulation in the plants. Such plants when they are fed to livestock can lead to toxicity. This has been recorded in the South Dakota area of USA, in some regions of Punjab in Pakistan, and also in North-West provinces of India. In India and Pakistan it is known by the disease "Degnala", whereas in USA it is called "Alkali Disease". This toxicity can occur in livestock as well as in human beings. In India it has been recorded in buffaloes and cattle, being more severe in buffaloes than in cattle. It is a chronic disease, which may take up to 42 days before symptoms appear.

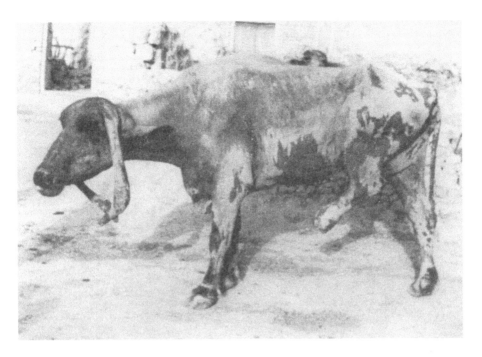

**Figure 3**. *A buffalo showing degenerated and sloughed off hoof.*

Syndromes usually appear after the monsoon season and the cases are recorded as late as May every year. Experimentally, the disease has been induced by feeding rice straw, rice husk and even berseem containing a higher level of Se. The animals show lesions on the tail, ear tips, limbs distal to knee joints and hind limbs below the hock joints. Its symptoms sometimes appear on the muzzle and back. The skin and hooves are the common tissues which are affected (Arora *et al.*, 1975). There is usually necrosis of tail and ear tips in the first instance, the legs show swelling followed by skin necrosis, skin desquamation, open wounds, and even gangrene in the affected region. In very severe cases the hooves slough off (Figure 3) and the animals may ultimately succumb. The disease has also been induced in pigs by Mahan and Moxon (1981) by feeding 10 ppm selenium, with manifestation of cracked hooves as a result of toxicity. The usual range of Se contents in different straws and fodders is from 0.9 to 6.7 ppm in samples collected from the farmers having affected cattle (Arora *et al.*, 1975). Though it is not clear as yet as to the minimum toxic level, it is apparent that the Se level and the level of protein in the total diet play an important role for the toxicity symptoms to develop. The higher the protein, the lesser the chances of Se toxicity. In rice straw and rice husk the quantity of digestible protein is negligible, and therefore it has been observed that the levels beyond 0.5 ppm may prove toxic. It has been estimated that the total absorption of Se from dietary sources is only 29%

**Figure 4.** *Syndromes of necrotic skin on lower region of the leg.*

(Khirwar and Arora, 1976). Se in the body tissues acts through its organic compounds, seleno-cysteine and seleno-methionine at the cellular level in the process of protein synthesis. It appears as if the cells which synthesise protein containing seleno-cysteine and seleno-methionine become necrosed (Figure 4). Both seleno-cysteine and seleno-methionine are present in milk in the casein fraction which can also affect neonatals. The animals suffering from Se toxicity do not lose appetite and they continue to eat their normal diet and some even continue to give milk (Khirwar and Arora, 1977). Selenium also inhibits the function of sulphydryl groups which are contained in lipoic acid, glutathione and even in some other components which are active in cellular respiration. Competition of Se with S in body tissues is because of its close relationship crystallo- chromically as well as geochemically. Selenium content in soils has been reported from different provinces of this country by Sharma et al., 1981. In Haryana, levels in soils have been reported to range from 1 ppm to 9.5 ppm. There are certain reports which indicate that certain crops accumulate more Se if S is deficient in the soil. Patil and Mehta (1970) observed that 2 ppm Se produced harmful effects on wheat plant but such effects were evaded by S application. Excess Se has been reported to inhibit the growth of guinea grass and jowar under the same conditions of Se level and S supplementation. Dhillon (1972) reported that the application of Se reduced the dry matter yield of maize in a field as compared to controls. Thus, as shown in the reports, 1-2 ppm Se is toxic to the plant which can be ameliorated by applying S. Prasad and Arora (1980) reported that Se accumulation in rice plant varies from 5.24 to 9.70 times in grain and 8.06 to 11.66 times in straw given doses ranging in level from 0.27 to 1.50 ppm. Both grain and straw yields decrease with an increase in Se dose up to 0.75 ppm.

*Induction of disease syndromes*

In order to establish the aetiology of Degnala disease, the results of certain experiments indicate that if buffalo calves are given parental intra-muscular administration of seleno- methionine (SeM) or SeM plus seleno-cysteine (SeC) in doses of 0.1 and 0.03 mg/kg live weight, the animals exhibit Degnala disease syndromes. The severity of symptoms in most of the animals is up to plus 2 out of 5 after a period of 8 to 18 days. The symptoms are usually of the type of skin necrosis mostly on the tail, ear tips and legs below the hock or knee joints. When SeM is injected subcutaneously at the root of tail, the severity of induced symptoms is of the degree of plus 4 out of 5 (Figure 5a,b) (Arora et al., 1987). It is evident from these experiments that organic sources of Se such as SeM and SeC present in certain straws and fodders may be the actual cause of this toxicity, namely Degnala disease in buffaloes and cattle.

*Line of treatment*

It has been reported that in the body tissues there is degradation of L-seleno-cysteine in the presence of enzyme (seleno-cysteine β-lyase) + 2 RSH or 2 RSEH into L-alanine + $H_2$ Se (Soda et al., 1981; Esaki et al., 1982); seleno-cysteine β- lyase occurs in various mammalian tissues such as liver and kidneys having higher activity than those of other tissues such as pancreas, adrenal, heart, lung, testis, brain thymus, spleen and muscle of rat, dog, pig and some other mammals. No significant activity is detected in rat blood or rat fat. Seleno- cysteine and seleno-methionine are fully active for the enzymes of S metabolism due to which some tissue proteins are altered by substitution of methionine by seleno-methionine. Selenium is incorporated into the proteins at the disulphide bridge to form seleno-trisulphite at low pH whereas at pH 7 the reaction leads to extensive liberation of element Se. Keeping in mind this approach, sulphate treatment has been devised to treat Se toxicity cases in chronic selenosis. In the urine Se is excreted in the form of trimethionine selenide (Palmer et al., 1969) by the probable action of sulphate. This therapy as "Degcure mixture" has helped to alleviate the toxicity symptoms in induced experimental animals (Arora, 1985).

**Figure 5 (a,b).** *Induced syndromes of Degnala disease by injecting SeM..*

# References

Ahuja, A.K. and Arora, S.P. (1980). Sulphur requirements of growing and lactating crossbred cows (Brown Swiss x Sahiwal). *Z. Tierphys Tierernahrg. U. Futtermittelkde*, **44**, 190-197.

Ahuja, A.K. and Arora, S.P. (1981). Utilisation of nitrogen and sulphur by rumen microbes and their excretion in calves. *Indian J. Anim. Sci.*, **51**, 1128-1132.

Arora, S.P. (1981). Zinc and vitamin A interrelationship in metabolism. *TEMA-4, May 11-15, Perth (Australia)* pp.572-574.

Arora, S.P., Ludri, R.S., Singhal, K.K. and Atreja, P.P. (1974). Ammonium sulphate as a nitrogen and sulphur source for utilisation by rumen microbes. *Indian J. Dairy Sci.*, **27**, 300-303.

Arora, S.P., Parvinder Kaur, Khirwar, S.S., Chopra, R.C. and Ludri, R.S. (1975). Selenium levels in fodders and its relationship with Degnala disease. *Indian J. Dairy Sci.*, **28**, 49-253.

Arora, S.P. (1985). *Proc. 1st International Symposium on Geochemistry and Health*, Science Reviews Ltd., Northwood, Middlesex, HA6 3DY.

Arora, S.P., Prasad, T. and Chopra, R.C. (1987). Seleno- methionine as an aetiology of Degnala disease in bovines. *Indian J. Animal Nutr.* (In press).

Bremner, I. (1981). The nature and function of metallothionein. *TEMA-4, Perth, Australia*, p.94.

Chhabra, A., Arora, S.P. and Jai Kishan (1980). Note on the effect of dietary zinc on $\beta$-carotene conversion to vitamin A. *Indian J. Anim. Sci.*, **50**, 879-881.

Chhabra, A. and Arora, S.P. (1985). Effect of zinc deficiency on serum vitamin A level, tissue enzymes and histological alterations in goats. *Livestock Prod. Sci.*, **12**, 69-77.

Danks, D.M. (1981). Trace elements in human disease with particular reference to copper and zinc. *TEMA-4, Perth, Australia*, p.79.

Dhillon, K.C. (1972). Selenium status of common fodders and natural grasses. M.Sc. Thesis, Punjab Agricultural University, Ludhiana.

Esaki, N., Nakumura, T., Tanaka, H., Soda, K. (1982). Seleno- cysteine $\beta$-lyase, a novel enzyme that specifically acts on seleno-cysteine. *J. Biol. Chem.*, **257**, 4386.

Gupta, B.S., Arora, S.P., Richaria, V.S., Ranjhan, S.K., Dhodapkar, B.S., Vagad, J.L. and Byers, J.F. (1972). Dietary investigation of causes of suspected parakeratosis in swine. *Indian J. Hlth.*, **11**, 65-67.

Hesketh, J.E., Watts, A. and Arbuthnott, G.W. (1985). A radioisotope and HPLC study of catecholamine sythesis in copper deficiency. *TEMA-5*, p.32. Aberdeen, UK.

Howell, J.M.C., Pass, D.A. and Terlecki, S. (1981). 4th International Symp. on Trace element metabolism in man and animals. *TEMA-4*, p.43. Perth, Australia.

Kanwar, J.S. and Randhawa, N.S. (1967). Micronutrient research in soil and plants in India. *A review, I.C.A.R.*, New Delhi.

Khirwar, S.S. and Arora, S.P. (1976). Influence of different levels of selenium on protein synthesis by rumen microbes *in vitro*. *Milchwissenschaft*, **31**, 275-77.

Khirwar, S.S. and Arora, S.P. (1977). Incorporation of Se-75- seleno-methionine in milk proteins of goats. *Milchwissenschaft*, **32**, 283-85.

Kraus, R.J., Foster, S.J. and Ganther, N.E. (1983). Identification of seleno-cysteine in glutathione peroxidase by mass spectroscopy. *Biochemistry*, **22**, 5853.

Kumar, S.K., Rowse, C., Hockstein, P. (1978). Copper-induced generation superoxide in human red cell membrane. *Biochem. Biophys. Res. Comm.*, **83**, 587.

Lal, B.M., Suba Rao, D. and Dase, N.B. (1960). *Current Science*, **19**, 250.

Mahan, D.C. and Moxon, A.L. (1981). Se and vitamin E for weanling swine. *TEMA-4*, p.207. Perth, Australia.

Merck, A.S. and Dev, G. (1979). Response of sunflower (*Helianthus annus*) to S application and evaluation of sulphur status of soils. *J. Nuclear Agri. & Biol.*, **8**, 100-102.

Mills, C.P., Dalgarno, A.C., Wenham, G. (1976). Biochemical and pathological changes in tissues of Friesian cattle during the experimental induction of copper deficiencies. *Br. J. Nutr.*, **35**, 309.

Neathery, M.W., Miller, W.J., Blackman, D.M. and Gentry, R.P. (1973). Performance and milk zinc from low zinc intake in Holstein cows. *J. Dairy Sci.*, **56**, 212.

O'Dell, B.L., Hardwick, B.C., Reynolds, G., Savage, J.E. (1961). Connective tissue defect in the chick resulting from copper deficiency. *Proc. Soc. Expt. Biol. Med.*, **108**, 402-405.

Ott, E.A., Smith, W.H., Martin, Stob and Beason, W.M. (1965). Zinc deficiency syndrome in the young lamb. *J. Nutr.*, **82**, 41.

Palmer, I.A., Cunsalue, R.P., Halverson, A.K. and Dison, D.E. (1969). Trimethylselenium ion as a general excretory product from selenium metabolism in the rat. *Biochem. Biophys. Acta.*, **208**, 260.

Patil, C.A. and Mehta, B.V. (1970). Se status of soil and common fodders in Gujarat. *Indian J. Agr. Sci.*, **40**, 389-99.

Prasad, C.S. and Arora, S.P. (1978). Influence of dietary zinc on the activity of alcohol dehydrogenase in different tissues. *Indian J. Anim. Sci.*, **48**, 582-84.

Prasad, C.S. and Arora, S.P. (1979). Influence of dietary zinc on β-carotene conversion and on the level of retinol binding protein in the blood serum. *Indian J. Dairy Sci.*, **32**, 275-79.

Prasad, T. and Arora, S.P. (1980). [75]Se-accumulation in rice plant and its effect on yield. *J. Nucl. Agri. Biol.*, **9**, 77-78.

Prasad, T., Arora, S.P. and Behra, G.D. (1982). Note on dietary investigation of suspected swayback in kids. *Indian J. Anim. Sci.*, **52**, 837-40.

Rifkind, J.M. (1974). Copper and autoxidation of haemoglobin. *Biochemistry*, **13**, 2474.

Saraswat, R.C. and Arora, S.P. (1972). Zinc deficiency syndromes in lambs. *Indian Vet. J.*, **49**, 701-04.

Sharma, S.K. and Prasad, T. (1983). Blood copper, iron, magnesium and zinc levels in growing and lactating buffaloes. *Indian J. Dairy Sci.*, **35**, 209.

Sharma, S. Ramendra Singh and Bhattacharya, A.H. (1981). Perspectives of selenium research in soil plant animal system in India. *Fertiliser News*, **26**, 19-28.

Sharma, V. (1982). Blood SOD activity and [75]Se distribution at different levels of dietary Cu in goats. M.Sc. Thesis, Kurukshetra University, India.

Smith, R.M., King, R.A., Osborne-White, W.S. and Fraser, F.J. (1981). Copper and the pathogenesis of enzootic ataxia. *TEMA-4*, p.42. Perth, Australia.

Soda, K., Nakamura, T., Tannaka, H. and Esaki, N. (1981). Mammalian synthesis and degradation of seleno-cysteine. *TEMA-4*, p.625. Perth, Australia.

Sonawane, S.N. and Arora, S.P. (1975). Influence of dietary levels of zinc on its secretion in milk. *Indian J. Dairy Sci.*, **38**, 99-103.

Underwood, E.J. (1977). In: *Trace Elements in Human and Animal Nutrition*. 4th edn. Academic Press, New York.

Whagner, P.D., Oh, S.H. and Deagen, J.T. (1981). Effects of dietary sulphur on tissue metallothionein in rats. *TEMA-4*, p.95. Perth, Australia.

# 14 Trace Elements and Stable Isotopes in Prehistoric Human Skeletons

Gisela Grupe
*Institut für Anthropologie der Georg-August-Universität, Bürgerstrasse 50, 3400 Göttingen, FRG*

## Summary

*Trace element and stable isotope analysis of prehistoric human hard tissue remains (mainly bones, but also teeth and hair) is a means of reconstructing ancient dietary habits and of the evaluation of certain pathological features in ancient societies.*

## Introduction

The department of "Prehistoric Anthropology and Environmental History" of the Institute of Anthropology, University of Göttingen (FRG), has recently started the research project "Trace Elements in Environmental History" (Grupe and Herrmann, 1987). The aim of this project is the reconstruction of the relations between man and his environment in the past. Knowledge of ancient human lifestyle, subsistence strategies and health hazards can be very helpful in understanding modern conditions. Research subjects are the remains of the people themselves, mainly excavated skeletons.

A very promising new field of research is the analysis of trace elements and stable carbon and nitrogen isotopes in ancient human remains. Many of the elements entering the organism and its metabolic cycles are stored in the skeleton. Since bone has a relatively slow turn-over rate, its composition reflects the long-term intake of an element in question. Analysis of the elemental composition of skeletons therefore gives direct information on the chemical constituents of the environment. This holds true especially for prehistoric populations, since both migration and food transport covering long distances were very limited.

Changes of the bone's elemental composition may occur during interment of the corpse and may cause problems. However, isotopic fractionation with regard to C and N apparently does not occur in the bone's collagen (Nelson *et al.*, 1986). As far as the bone mineral is concerned, the concentrations of those elements firmly bound to the hydroxyapatite (like Sr, Zn, and Pb) are not, or are only slightly, affected by diagenesis. Other elements like Al, Fe, and Mn are contaminating the skeletons, thus reflecting soil features rather than *in vivo*-uptake (for summary see Grupe, 1986).

## Methods

Compact bone specimens are recommended for analysis. Stable isotope analysis requires the extraction of collagen by acid hydrolysis. The remaining gelatinous fraction is combusted and the isotopic ratios ($\delta^{13}C$, $\delta^{15}N$) determined by mass spectrometry. For trace element analysis sample processing depends on the measuring technique (AAS, ICP, NAA). In every case cleaning procedures should include ether-extraction of adherent organic particles followed by ultrasonic cleaning with formic acid (for details see Grupe, 1986). In order to check whether bone decomposition has caused trace element loss or contamination soil samples should always be taken according to forensic exhumation techniques to analyse the soil trace element content as well.

## Diagnostic Potential and Perspectives

### *Reconstruction of ancient food chains*

The chemical constituents of the body tissues enter the organism mainly by intake of food and drinking water. Many elements including stable C- and N-isotopic ratios show significantly different concentrations in the basic food items which are maintained throughout the food chain. Trace element contents and stable isotope ratios in excavated human skeletons therefore lead to the reconstruction of ancient dietary habits.

In general, the following estimations concerning the composition of daily diet in the past are possible:

- The relative amount of food of vegetal or animal origin. For example, plant food is enriched with Sr but poor in Zn, while meat, milk and dairy products contain much Zn, but are poor in Sr. Thus, the Sr:Zn-ratio in the bone is a good estimate for the amount of plants versus meat and dairy products consumed.
- The relative amount of seafood. Compared with the terrestrial environment, sea water is enriched with $^{13}C$ and $^{15}N$. Thus, the tissues of people consuming large amounts of seafood are also enriched with both the heavy isotopes.
- The relative amount of certain plant groups. Most important for central European ecosystems is the differentiation between leguminous and non-leguminous plants by $\delta^{15}N$.

This information provide insights into the quality of ancient diets, into age- and sex-specific dietary habits, and into changes of subsistence strategies in prehistoric populations (*e.g.* hunting and gathering *versus* agriculture).

Moreover, it is possible to estimate the average weaning age in a given population by trace element analysis of the young infants' bones, since weaning means a shift in a baby's diet from mother's milk to diets containing a higher amount of plant food. Knowledge of weaning patterns in correlation with the mortality pattern among infants give clues to parental investment and reproductive strategies in the past.

*Pathology*

Environmental diseases, like heavy metal intoxication, are reflected in the trace element contents in the skeleton. The lead burden of ancient societies, for example, has frequently been studied (*e.g.* Waldron and Wells, 1979).

Trace element analysis also leads to the reconstruction of individual health hazards. Recently, a medieval skeleton (male, 40-50 years old; 11th/12th century) was excavated in northern Germany, whose bone lesions indicated that the individual had suffered from a bronchogenetic carcinoma (Grupe, 1986). The skeleton's trace element composition did not differ from those of the other adult individuals on the graveyard (total: 87 adult skeletons) except with regard to Sb. The bones of the diseased person contained as much as 13.940 ppm Sb compared with a mean Sb-content of 0.186 ppm in the other skeletons. This indicates a very high exposure to Sb of this particular individual whose occupation most probably was some kind of metal working. Thus this skeleton represents an early case of an occupational disease.

It is most likely that certain metabolic disorders can also be detected by trace element analysis of excavated human skeletons. Unfortunately, the relation between the trace element content in body fluids and bone in diseased persons is often unknown. This opens up a new field for future research.

Parts of this work have been supported by the Deutsche Forschungsgemeinschaft.

## References

Grupe, G. (1986). Multielementanalyse: Ein neuer Weg für die Paläodemographie. *Materialien zur Bevölkerungswissenschaft.* Sonderheft, 7. Wiesbaden.

Grupe, G. and Herrmann, B. (1987). *Trace Elements in Environmental History.* Springer, Berlin, Heidelberg, New York (in press).

Nelson, B., DeNiro, M., Schoeninger, M., DePaolo, D. and Hare, P. (1986). Effects of diagenesis on strontium, carbon, nitrogen and oxygen concentration and isotopic composition of bone. *Geochim. Cosmochim. Acta*, **50**, 1941-1949.

Waldron, T. and Wells, C. (1979). Exposure to lead in ancient populations. *Trans. Stud. Coll. Phys.*, Philadelphia, Ser. V., Vol.1, 102-115.

# 15 Lead and Other Metals in the Compartments of a Contaminated Ecosystem in Wales

M.R. Matthews and B.E. Davies[*]
[*] *School of Environmental Sciences, University of Bradford,*
*West Yorkshire BD7 1DP, England*

## Abstract

*This paper describes the results of soil metal surveys in a contaminated and an uncontaminated catchment in Wales. Summary data are presented for the total concentrations of Pb, Zn, Cu, Cd, Co, Ni, Mn and Fe. The mean metal concentrations of each compartment are also given, together with Pb loadings. The data are discussed in terms of long-term pollution hazard and target compartments in the contaminated catchments and compared with the distributions of metals in the control catchment.*

## Introduction

Great Britain was once among the major world producers of non- ferrous metal ores but, towards the end of the nineteenth century, the industry declined sharply in the face of foreign competition. At the present time, metal ore output in this country is confined to tin in Cornwall (Richardson, 1974), and to lead and zinc (as by-products from fluorspar mining) in Wales and Derbyshire (Borough of Wrexham Maelor *et al.*, 1974).

The dereliction produced by lead mining in Wales has left a legacy of widespread heavy metal contamination. Metal-rich waste was dumped anywhere convenient around the mines, and these wastes are subject to remobilisation so that the toxic metals are transported to "sinks", for example, soils, sediments and biota. The processes involved in the transfer of metals have been investigated by many workers (*e.g.*, Lewin, *et al.*, 1977; Johnson and Roberts, 1978; Davies, 1971a) and it is generally accepted that there are three main processes involved, namely, fluvial, atmospheric and gravitational transportation. These have generated a complex pattern of soil contamination which has been well described and is understood in general terms.

In contrast, little is known about the detailed distribution of metals in linked compartments of an ecosystem. This is an important prerequisite to modelling movement and hence predicting the long-term fate of metals. Thus, the primary objective of the project was to provide such information using a small contaminated catchment.

## Study Areas

Cwmsymlog and Cwmerfin are two derelict Pb/Ag/Zn mines in a small, mixed land use catchment in Dyfed, Wales. Previous research (Davies and White, 1981) described how loss of metal- rich materials from the waste heaps of these mines had led to widespread contamination by lead and other metals. A similar uncontaminated ecosystem was chosen for study at Cwn Nant Y Bompren, in North Powys.

## Materials and Methods

Topographical maps, air photographs and ground surveys were used to identify, classify and map the present land-use of the catchments. From these maps, areas of grassland, coniferous and deciduous woodland, marshland, alluvial soils, sediments and mine wastes were estimated. Detailed soil sampling was then undertaken using a 200 x 200 m grid in each catchment. At each sampling point, the soil mantle depth was measured, the land-use noted and a soil sample (0-15 cm) taken. Metal analysis was carried out by flame atomic absorption spectrometry, after extraction with hot

**Table 1.** *Compartment areas for both catchments based on published O.S. maps, aerial photographs and field surveys. Areas calculated on a digitising table.*

| Compartment | Catchment 1 Cwmsymlog Area ($m^2$) | % | Catchment 2 Cwm Nant Y Bompren Area ($m^2$) | % |
|---|---|---|---|---|
| Alluvial soils | 15,000 | 5.5 | 3,800 | 1.4 |
| Con. woodland | 47,500 | 16.3 | 13,700 | 5.2 |
| Dec. woodland | 11,400 | 3.9 | 36,300 | 13.9 |
| Mix. woodland | 200 | 0.1 | - | - |
| Grassland | 187,900 | 64.6 | 209,800 | 80.4 |
| Lakes/ponds | 4,700 | 1.6 | 200 | 0.1 |
| Marsh | 2,300 | 0.8 | 100 | 0.01 |
| Mine waste | 31,900 | 10.9 | - | - |
| Non-soil seds. | 4,900 | 1.7 | 1,000 | 0.4 |
| Total area | 290,900 | | 261,100 | |

Table 2. *Summary statistics for topsoils in Cwmsymlog (A) and Cwm Nant Y Brompren (B). Spring, 1985. Metals expressed as mg/kg, except Fe (expressed as a percentage). Soil reaction (pH) determined after equilibration with 0.01M CaCl₂. Organic matter expressed as a percentage after loss on ignition at 430°C.*

| | Mean | Median | Max | Min | SD | Geometric mean | SD |
|---|---|---|---|---|---|---|---|
| Pb A | 2,896 | 100 | 81,818 | 41 | 11,170 | 259 | 6.2 |
| B | 31 | 28 | 185 | 6 | 20 | 28 | 1.5 |
| Zn A | 264 | 43 | 4,167 | 12 | 658 | 69 | 4.0 |
| B | 59 | 52 | 534 | 25 | 57 | 53 | 1.4 |
| Cu A | 69 | 7 | 3,421 | 2 | 384 | 12 | 3.4 |
| B | 9 | 8 | 38 | 2 | 5 | 8 | 1.6 |
| Cd A | 1 | 0.6 | 5.2 | 0.3 | 0.9 | 0.8 | 1.9 |
| B | 0.5 | 0.5 | 2.1 | 0.3 | 0.2 | 0.5 | 1.3 |
| Co A | 9 | 8 | 43 | 3 | 5 | 8 | 1.4 |
| B | 8 | 8 | 17 | 3 | 2 | 8 | 1.3 |
| Ni A | 11 | 10 | 64 | 4 | 7 | 9 | 1.4 |
| B | 11 | 10 | 38 | 5 | 5 | 10 | 1.4 |
| Mn A | 424 | 414 | 874 | 22 | 165 | 377 | 1.7 |
| B | 307 | 312 | 691 | 50 | 118 | 285 | 1.5 |
| Fe A | 1.4 | 0.2 | 3.5 | 0.01 | 0.8 | 0.2 | 4.8 |
| B | 1.3 | 1.4 | 2.3 | 0.004 | 0.5 | 0.9 | 3.8 |
| pH A | 6.0 | 6.1 | 6.6 | 3.5 | 1.2 | 5.8 | 1.1 |
| B | 5.2 | 5.4 | 6.3 | 4.2 | 0.8 | 5.0 | 1.1 |
| %OM A | 13 | 13 | 75 | 0.6 | 9 | 11 | 2.0 |
| B | 11 | 11 | 20 | 5 | 3 | 11 | 1.2 |

*aqua regia* (3:1; nitric:hydrochloric acid). pH was measured after equilibration with 0.01 M calcium chloride; % organic matter was estimated by the loss on ignition at 430°C.

## Results and Discussion

Compartment areas are presented in Table 1. Both catchments are of a similar size and land-use pattern, apart from the absence of mine waste in the control catchment. One other notable difference is the dominance of coniferous woodland in Cwmsymlog whereas deciduous trees dominate Cwm Nant Y Bompren. This is probably due to deforestation in the early years of mining since wood was used for smelting on site (Lewis, 1967). The agriculturally unproductive slope soils have since been planted with conifers. The native deciduous woodlands have continued to flourish, however, in Cwm Nant Y Bompren.

**Table 3.** *Mean metal concentrations in topsoils of ecosystem compartments. Cwmsymlog and Cwnerfin (A), Cwn Nant Y Bompren (B).*

| | | Alluvial soils | Con. wood | Dec. wood | Grass | Marsh | Mine waste | Non-soil sediments |
|---|---|---|---|---|---|---|---|---|
| Pb | A | 22,190 | 143 | 93 | 405 | 512 | 4,366 | 915 |
| | B | 43 | 7 | 27 | 24 | 42 | - | 184 |
| Zn | A | 639 | 44 | 40 | 75 | 44 | 1,087 | 51 |
| | B | 62 | 33 | 45 | 60 | 45 | - | 534 |
| Cu | A | 63 | 12 | 7 | 10 | 12 | 367 | 13 |
| | B | 12 | 4 | 8 | 9 | 10 | - | 38 |
| Cd | A | 0.8 | 0.7 | 0.8 | 0.7 | 2.9 | 3.1 | 1.0 |
| | B | 0.6 | 0.4 | 0.5 | 0.5 | 0.7 | - | 2.1 |

The summary statistics for total metals are outlined in Table 2. The most striking feature for the Cwmsymlog data is the tendency for the arithmetic mean to be greater than the median. This implies positive skewness, *i.e.* the population is abnormally influenced by high values, suggesting contamination. This was overcome by a $\log_{10}$ transformation of the data. Metal values for the control catchment are much lower and are consistent with the background levels obtained by Davies (1987) for Wales.

Examination of the ecosystem compartments (Table 3) shows mine waste and alluvial soils to have highest metal levels in the contaminated catchment and non-soil sediments are richest in metals in Cwm Nant Y Brompren. It may be suggested that mine waste is the major "source" of metals and alluvial soils the major "sink". Lowest metal concentrations were obtained for coniferous woodland soil in both catchments. This is probably due to enhanced leaching down the profile in the process of podsolisation (Davies, 1971b).

However, these values, although representing the average metal concentration of soils in each compartment, do not give any indication of the total metal mass or loading. Using the soil mantle depth measurements, metal loadings were calculated for each catchment, and the results for lead are presented in Table 4. It may be noted that the greatest mass (2,247 g or 44%) is represented by mine waste and a further 39% by alluvial soils in the Cwmsymlog catchment. In the control catchment, 81 g (or 92%) is represented by grassland. A further 7% was in deciduous woodland. When the lead data were weighted according to compartment size, unusually high loadings were found for alluvial soils and mine waste in the contaminated catchment and in grassland in the control catchment.

**Conclusions**

It is concluded that mine waste is a major source of metals in Cwmsymlog and Cwmerfin. The major primary sink is alluvial soils, followed by grassland. There are

**Table 4.** *Lead loadings and weightings in compartments of Cwmsymlog (A) and Cwm Nant Y Bompren (B).*

| Compartment | | Area | Pb loading (g) | (%) | Pb weighting |
|---|---|---|---|---|---|
| Alluvial | A | 15,900 | 2,037 | 39.800 | 7.200 |
| soils | B | 3,800 | 0.1 | 0.100 | 0.070 |
| Coniferous | A | 47,500 | 29 | 0.600 | 0.040 |
| woodland | B | 13,700 | 0.5 | 0.500 | 0.310 |
| Deciduous | A | 11,400 | 1.0 | 0.020 | 0.005 |
| woodland | B | 36,300 | 6.0 | 7.000 | 0.560 |
| Grassland | A | 187,900 | 797 | 15.600 | 0.260 |
| | B | 209,800 | 81 | 92.100 | 1.200 |
| Marsh | A | 2,300 | 2.0 | 0.030 | 0.040 |
| | B | 100 | 0.003 | 0.003 | 0.070 |
| Mine waste | A | 31,900 | 2,247 | 44.000 | 4.000 |
| Non-soil | A | 4,900 | 3.0 | 0.050 | 0.030 |
| sediments | B | 1,000 | 0.21 | 0.240 | 0.600 |

no major sources in the control catchment, although a rubbish tip may contribute as a minor source. The slightly elevated weighting in grassland is probably entirely due to the size of this compartment. Once metals reach soils they are likely to remain there for a long time. Lead and cadmium are especially concentrated in the upper horizons and are unlikely to be appreciably depleted by leaching (Korte, *et al.*, 1975). Work is in progress to estimate the rates of metal movement between compartments. The data will then be used to model the catchments and hence predict the ultimate fate of these metals.

### Acknowledgements

The authors are grateful to Powys Education Committee and Bradford University for financial support.

### References

Borough of Wrexham Maelor, Montgomery District Council, Department of Botany, Liverpool University and Robinson Jones Design Partnership Ltd. (1975). A Pilot study into the rehabilitation of metalliferous mine waste at Minera and Y Fan.

Davies, B.E. (1971a). Trace metal content of soils affected by base metal mining in the West of England. *Oikos*, **22**, 366-372.

Davies, B.E. (1987). Baseline survey of metals in Welsh soils. *Proceedings of the First International Symposium on Geochemistry and Health*, pp.45-51. Science Reviews Ltd, Northwood, UK.

Davies, B.E. and White, H.M. (1981). Environmental pollution by wind blown lead mine waste: a case study in Wales, UK. *Sci. Total Environ.*, 20, 57-74.

Davies, R.I. (1971). Relation of polyphenols to the decomposition of organic matter and to pedogenetic processes. *Soil Sci.*, 111, 80-85.

Johnson, M. and Roberts, D. (1978). Lead and zinc in the terrestrial environment around metalliferous mines in Wales (UK). *Sci. Total Environ.*, 10, 61-78.

Korte, N.E., Skopp, J., Niebla, E.E. and Fuller, W.H. (1975). A baseline study on the trace metal elution from diverse soil types. *Water, Air and Soil Pollution*, 5, 149-156.

Lewin, J., Davies, B.E. and Wolfenden, P.J. (1977). Interactions between channel changes and historic mining sediments. In: K.J. Gregory (ed.), *River Channel Changes*, pp.353-367. John Wiley and Sons Ltd., Chichester.

Lewis, W.J. (1967). *Lead Mining in Wales*. University of Wales Press.

Richardson, J.B. (1974). *Metal Mining*. Allen Lane, London.

# 16 Geochemistry and Health in Hungary

Irén Varsányi
*Köjál 6726 Szeged Derkovits Fasor 7-11, Hungary*

## Summary

*The differences in the regional distribution of mortality from cancer, cardiovascular, respiratory and other diseases can be explained by the regional distribution of drinking water quality. This connection can be pointed out in a relatively small area only by multivariate methods such as canonical correlation analysis and cluster analysis.*

## Introduction

There are a lot of factors influencing mortality. These factors can modify the effect of each other so it can hardly be imagined that a relationship can be shown between only one of these factors and one of the rates referring to mortality.

## Location and Methods

Csongrád county is a small area (4,280 km$^2$) in the southern part of the Great Hungarian Plain. The number of inhabitants is 135,989. There are five towns and 43 villages in this district. The way of life and the medical care differ in the towns and villages, so mortality has to be studied separately. In this paper only the mortality in the villages is surveyed. The data concerning the main causes of death - cancer, cardiovascular, respiratory and the sum of other mortality, were collected in five consecutive years from death certificates. These data were age-standardised and referred to 1,000 inhabitants. The effect of total hardness, $O_2$ consumption, total Fe and As, $Na^+$, $K^+$, $NH^+_4$, $Cl^-$ on the mortality were studied.

In each village there is a central water supplying plant based on deep (200-600 m) subsurface water.

To evaluate the data individual correlation, multiple regression canonical correlation (Kendall and Stuart, 1979, Mardia *et al.*, 1979) and cluster analysis (Le Maitre, 1982) were used.

**Table 1.** *Individual correlation.*

|  | Cancer | Cardiovascular | Respiratory | Others |
|---|---|---|---|---|
| Total hardness | -0.273 | -0.198 | -0.135 | 0.279 |
| $O_2$ consumption | 0.168 | 0.077 | -0.118 | -0.371+ |
| Total Fe | -0.141 | -0.127 | 0.003 | 0.037 |
| Total As | 0.158 | -0.112 | 0.015 | 0.192 |
| $Na^+$ | 0.240 | 0.090 | 0.038 | -0.276 |
| $K^+$ | -0.070 | -0.049 | -0.080 | -0.159 |
| $NH_4^+$ | -0.079 | 0.166 | 0.203 | -0.344+ |
| $Cl^-$ | 0.250 | -0.149 | -0.004 | -0.072 |

$r_{42;0.05} = 0.304$

**Table 2.** *Multiple correlation.*

|  | $X_1 - X_8$ | Correlation coefficient | F | DF |
|---|---|---|---|---|
| Cancer | Water Quality | 0.473 | 1.22 | 8 34 |
| Cardiovascular | " | 0.504 | 1.45 | 8 34 |
| Respiratory | " | 0.440 | 1.02 | 8 34 |
| Others | " | 0.512 | 1.51 | 8 34 |

$F_{8,34;0.05} = 2.23$

## Results

From the 32 possible individual correlation coefficients only two show a significant correlation between mortality and water quality (Table 1).

This does not mean a lack of any relation between water quality and mortality but only a weak correlation between the individual factors.

Multiple regression analyses were made for cancer, cardiovascular, respiratory and other mortality, respectively. The F test shows that even the multiple correlations are not significant (Table 2).

Significant correlation has been proved by canonical correlation analysis (Table 3). The first canonical correlation coefficient is significant and exceeds all the individual and multiple correlation coefficients.

**Table 3.** *Canonical correlation.*

| Eigen value | Correlation coefficient | F | DF |
|---|---|---|---|
| 0.397 | 0.630[+] | 46.21 | 32 |
| 0.355 | 0.595 | 27.73 | 21 |
| 0.150 | 0.388 | 11.75 | 12 |
| 0.147 | 0.384 | 5.81 | 5 |

$2_{32;0.05} = 43.8$

On the basis of the cluster analysis the study area can be divided into six groups considering water quality and eight groups considering mortality. The similarity of the regional distribution of both mortality and water quality is obvious.

**References**

Kendall, M., Stuart, A. (1979). *The Advanced Theory of Statistics*, Vol.2. Charles Griffin Co. Ltd.
Le Maitre, R.W. (1982). Developments in Petrology, *Numerical Petrology*. Elsevier Scientific Publishing Company.
Mardia, K.V., Kent, J.T., Bibby, J.M. (1979). *Multivariate Analysis*. Academic Press.

# 17 Hydrogeochemistry of Finnish Groundwaters - Some Implications for Health

Pertti W. Lahermo
*Geological Survey of Finland, SF-02150 Espoo, Finland*

## Summary

*The countrywide hydrogeochemical mapping of rural springs and wells carried out by the Geological Survey and the statutory quality monitoring of public water utilities by the National Board of Waters provide comprehensive data for evaluating the quality of Finnish groundwaters and their assumed health implications. Finnish groundwaters are very soft Ca(HCO3)2 waters, and concentrations are regulated by bedrock composition, aquifer structure and related groundwater regime and marine factors, i.e. secondary sulphur compounds in sediments and relict sea salts. The quality of water from private single household wells or supplied by public water utilities is generally good. Only contents of F, Rn, and NO3, with related bacteria, sometimes exceed the set standards and may cause health risks. The increasing acidification of the environment, however, presents a risk of raising Al and other metal contents in drinking water.*

## Introduction

In recent years more attention has been paid to the chemical quality of groundwaters for several reasons: increasing man- induced changes in groundwater used by private households and communities, the impact of acid rain on the environment and related health considerations and, in the case of deep bedrock groundwater, the planning of repositories for the disposal of nuclear waste in the bedrock. The Geological Survey of Finland carried out large-scale hydrogeological mapping of groundwater in rural areas to study the interrelationships between the geological and human environment and water quality. The comprehensive material collected provides a good opportunity for considering the quality of groundwater with reference to the geological environment and the health of the population (Table 1). An estimated 75-80% of the population is catered for by the public water utilities. Nearly half of the current water

169

**Table 1.** Chemical composition of water from springs, dug wells and drilled bedrock wells (data from the Geological Survey) and from public water utilities (Anon 1979, 1983, 1986). The well data are median and arithmetic mean values (median values are lower) and the water utility data are median values from 1977, 1980 and 1984.

| | EC mS/m | pH | CO$_2$ mg/L | KMnO$_4$ consm. | Hardn. dH° | HCO$_3$ mg/L | SO$_4$ mg/L | Cl mg/L | NO$_3$ mg/L | Fe mg/L |
|---|---|---|---|---|---|---|---|---|---|---|
| Springs (N ~ 1,200) | 7-11 | 6.2-6.3 | 72-74 | 5-8 | 1.3-1.9 | 21-29 | 5-9 | 2.5-6.8 | 0.8-4.6 | 0.1-0.2 |
| Dug wells (N ~ 3,500) | 19-25 | 6.4 | 59-60 | 8-14 | 3.3-4.2 | 46-65 | 13-19 | 8.9-17 | 5.6-14 | 0.1-0.4 |
| Drilled wells (N ~ 1,000) | 24-35 | 6.7 | 43-51 | 6-10 | 4.2-5.5 | 74-98 | 11-19 | 12.5-32 | 2.3-12 | 0.1-0.9 |
| Public water utilities | 12-16 | 6.5-6.7 | 16-20 | 2-4 | 2.0-3.2 | 24-43 | 15-28 | 7-14 | 0.9-1.5 | 0.04-0.1 |

(573/1977, 688/1980, 567/1984).

supply is groundwater. The quality is regularly monitored by law by the National Board of Waters and Environment at a sampling density and using analytical procedures compatible with the number of consumers. This is another large source of hydrochemical material for evaluating the quality of groundwater in this country.

**Groundwater Quality with Special Reference to the Geological Environment**

According to international standards Finnish groundwaters are very soft, being diluted $Ca(HCO_3)_2$ waters. Because the distribution patterns of Ca, $HCO_3$, total hardness and electrical conductance (EC) are very similar over the whole country, only a map of total hardness is given (Figure 1). The areas with the highest average contents are concentrated along the coast owing to the structure of aquifers and the marine effect. In coastal areas, the aquifers are often partly covered and bordered by confining clay and silt deposits. Therefore the amounts of dissolved components are somewhat higher than in open aquifers with a shorter groundwater retention time. There are, however, some interrelationships between water quality and bedrock composition, too. The wedge-shaped area to the north of the Bothnian Bay with elevated abundances of dissolved constituents in groundwater coincides with the Peräpohja schist area, which is composed, among others, of gabbros, anorthosites, ultrabasic rocks, dolomites and limestones (Figure 2). The boundary between the schist belt in central Lapland and the northern granulite and Archean granitoid area is clearly seen in groundwater quality (see the dashed line in Figure 2A). The large Archean basement in eastern Finland is likewise a region of low concentration groundwater.

The distribution of $SO_4$ in groundwater follows more or less the same trend as above, although the marine effect, with the elevated sulphate concentrations along the coast, is a more prominent feature (Figure 3). Sulphates may derive from relict sea salts left in marine clay and silt deposits and fractures of bedrock or from secondary sulphide and sulphate precipitates of marine sediments. In eastern and southern parts of the country there are also elevated $SO_4$ concentrations in the Outokumpu and Pori - Vammala - Tampere sulphide-ore zones (*cf.* Figure 2).

Fluorides exhibit the most striking interrelationship between water chemistry and bedrock composition, as the contents are many times higher in areas composed of rapakivi granites and related anorogenic granites than elsewhere in the country (Figure 4). Sporadically higher F contents in water also occur in other granitoid areas (especially coarse-grained K granites) in central and western Finland and in Lapland. Somewhat higher F contents in water continue to the southeast of the western rapakivi massif, partly because of glacial transport of the soil material in the same direction, and partly because of the occurrence of fluoride- rich minerals in granites, limestones and other rocks.

Iron and manganese contents are also at their highest level in low-lying coastal areas. Their occurrence is governed mainly by confined aquifer conditions, which favour oxygen deficiency, and by the occurrence of dissolved humic material with the ability to form compounds with heavy metals.

171

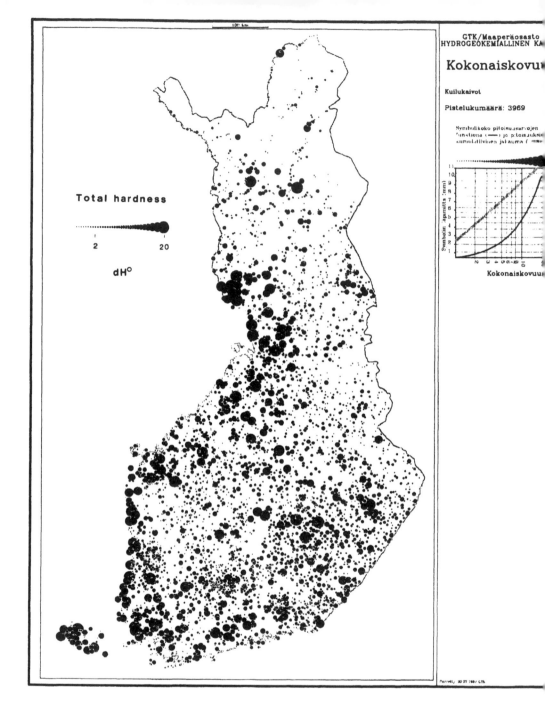

**Figure 1.** *Distribution of total hardness values in water from wells dug into overburden mainly composed of glacial till and, in c. 20% of cases, of glaciofluvial sand and gravel (3,969 sampling sites). Values are in German degrees of hardness (1 dH° is equivalent to 10 mg/L of CaO and MgO).*

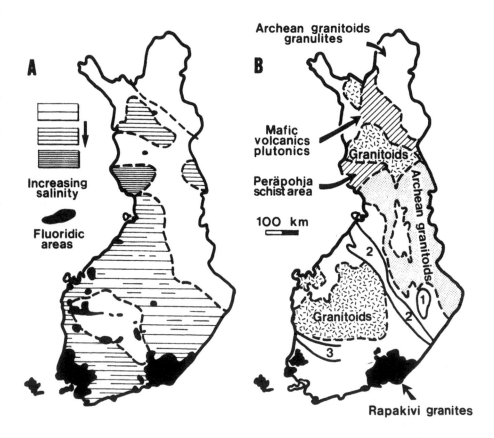

**Figure 2.** A: *Schematic salinity distribution and areas of highest fluoride values of dug well waters. The hydrogeochemical patterns drawn on the basis of specific conductance, total hardness, alkalinity and fluoride maps. B: The main features of the Pre-Cambrian bedrock (after Simonen, 1980). 1. The Outokumpu sulphide ore belt. 2. The Laatokka-Perämeri (Ladoga-Bothnian Bay) sulphide ore belt. 3. The Pori-Vammala-Tampere sulphide ore belt (after Kahma, 1973).*

Uranium contents in shallow groundwater are generally below the detection limit. For example, in the hydrogeochemical mapping of the Geological Survey only 4% of a total of more than 4,000 samples collected from shallow dug wells in the overburden gave results exceeding 1 μg/L. In spring waters the contents are still lower. Of the samples of bedrock groundwater taken from drilled wells 57% gave concentrations higher than 1 μg/L. The highest concentrations in bedrock groundwater are found in a broad belt across southern Finland (Figure 5A and B). The anomalous area is composed of granites or schists and migmatites intermingled with granitic veins. An uranium-anomalous zone coincides roughly with the radon-anomalous areas (Figure 5C). Radon emanates from the bedrock into drilled wells and the soil and thus via the basement into dwellings and other buildings.

173

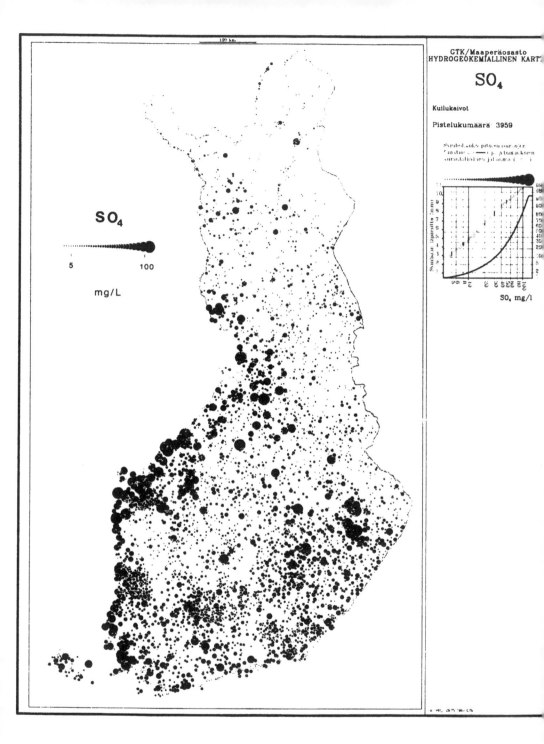

**Figure 3.** *Distribution of SO4 contents in water from wells dug into overburden (3,959 sampling sites).*

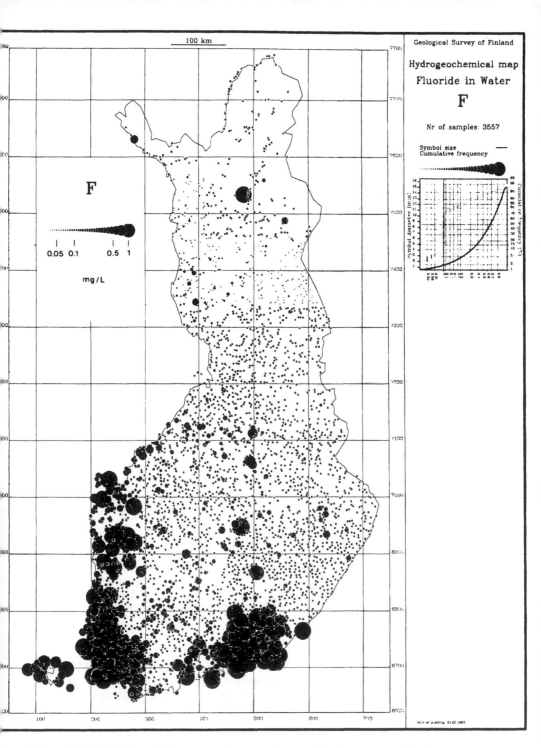

**Figure 4.** *Distribution of F contents in water from wells dug into overburden (3,557 sampling sites).*

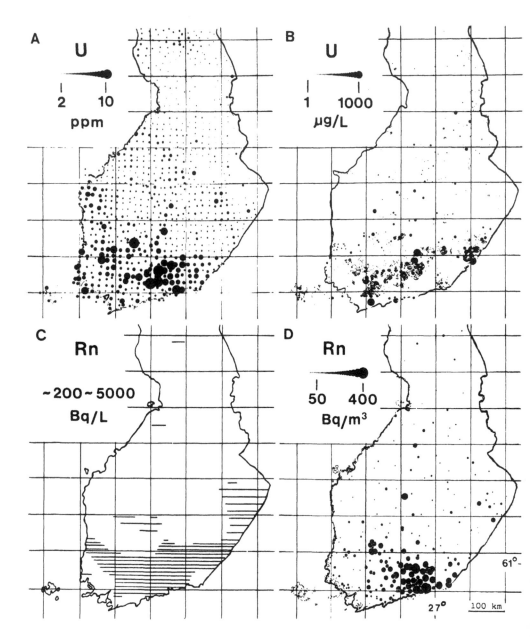

**Figure 5.** A: *Distribution of U contents in glacial till (1,056 sampling sites). B: Distribution of U contents in water from wells drilled into bedrock (2,370 sampling sites). C: Rough areal estimate of Rn contents mainly in bedrock groundwater, based on unpublished data from J. Hyyppä and from Asikainen and Kahlos (1977). In the more densely hatched area Rn concentrations may reach values of 5,000 Bq/L (the highest concentration so far met is 44,400 Bq/L). D: Distribution of Rn in indoor air (8,149 dwellings in 235 localities). Data from the Institute of Radiation Protection.*

176

Therefore, the highest Rn concentrations in indoor air occur in the same areas (Figure 5D).

The most effective and sensitive indicators of pollution are the $NO_3$ and K contents of water (Lahermo, 1987). Local contamination depends on the proximity of polluting sources, such as leaking sewage pipes or septic tanks, animal dung containers or heaps and waste tips. Another very important cause is deficient construction or condition of wells. The most highly contaminated rural dug wells are in central and southeastern parts of the country (Figure 6) although contaminated wells also occur in sparsely populated northern Finland. The elevated concentration in some parts of the country seems to be due to poor protection of wells in small farms engaging in animal husbandry. Traditionally, wells have been dug near cow sheds, making it possible for polluted run off to reach deficiently protected wells. Analyses from natural springs and drilled bedrock wells show sporadically elevated $NO_3$ contents all over the country.

## Groundwater Quality and Health

Public water utilities are responsible for the water supply of an estimated 75-80% of the population in Finland. The proportion of groundwater is roughly 50% and slowly increasing. The quality of drinking water is monitored chemically and bacteriologically and is generally good. In 442 and 688 groundwater intakes measured in 1977 and 1980, respectively, F contents exceeded the set standard (1.5 mg/L) in 8 to 11% of cases; Fe and Mn (1.0 and 0.5 mg/L) in 11 to 12% and 6 to 9% of cases; $NO_3$ and $NH_4$ (30 and 1.0 mg/L) in 1% of cases; and bacterial quality (more than 1,000 counts of coli/100 ml) in 2 to 3% of cases, respectively (Anon, 1979, 1983). Consequently, the quality of drinking water may be a cause for concern in only a few cases. Since the concentration of fluoride does not generally exceed 3-5 mg/L it cannot be considered a health risk under northern climatic conditions, where people do not drink large amounts of water. Iron and manganese cause only aesthetic and technical problems. The $NO_3$ contents are also generally low, although sporadically higher contents (more than 30 mg/L) as well as high indicator bacteria concentrations refer to local point-source anthropogenic contamination.

The quality of the other half of the water supplied by the public water utilities is roughly the same as that reported above with respect to groundwater. Surface water, however, is more acidic and coloured by humic matter, and bacterial contamination is slightly more common than in groundwater (the standard value was exceeded in 8% of cases).

The hydrogeochemical mapping project conducted by the Geological Survey on the groundwater occurring in overburden and bedrock in rural areas without access to reticulated public water supplies indicates greater variations in water quality. Higher F contents (1 to 2 mg/L) are encountered in only one part of the country (Figure 4), where they can be considered beneficial for humans. It has been found that the incidence of dental caries is lower in fluoridic areas than elsewhere. No harm is known to have been caused by elevated F contents in drinking water. Although the

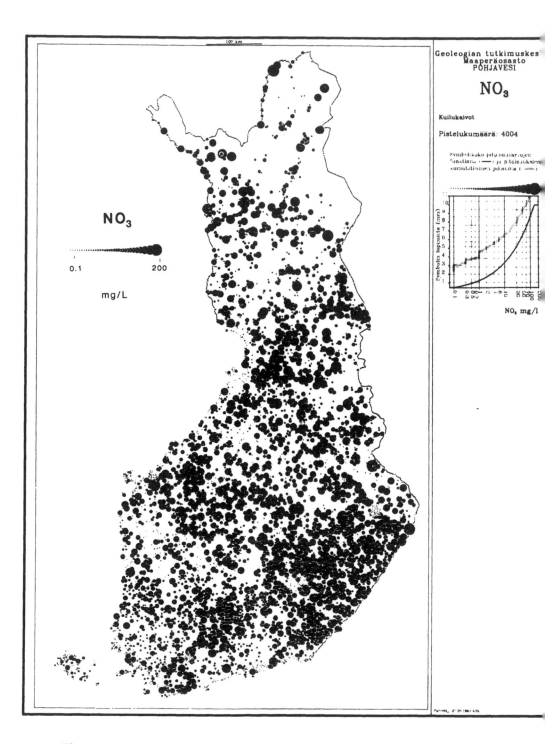

**Figure 6.** *Distribution of NO₃ contents in water from wells dug into overburden (4,004 sampling sites).*

F contents in water are too low in most of the country, only one town (Kuopio) adds fluorides to drinking water (up to 1 mg/L).

The hydrogeochemical mapping revealed that outside the western coastal belt iron and manganese do not create any marked problem, as Fe and Mn concentrations only seldom exceed 1.0 and 0.3 mg/L, respectively. Although severe anthropogenic pollution in individual rural wells is not common, the sporadically high $NO_3$ contents are a matter of concern in rural areas. Contents of $NO_3$ exceeded 32 mg/L in 3% of sampled springs (1,301 sites), in 9% of dug wells (3,018 sites) and in 6% of drilled bedrock wells (1,022 sites). The wells dug into glacial till deposits were more contaminated (10% of cases) than wells in more pervious sand and gravel deposits (contents of $NO_3$ exceeded 32 mg/L in 6% of cases).

Aluminium seems to occur in comparatively high concentrations in acidic shallow groundwaters. In the pH range of 4-5.5, Al contents may reach 1-3 mg/L. Aluminium is known to be harmful to fish in surface waters at concentrations of 0.5-1.0 mg/L, but there is no information about the effect of elevated Al contents in drinking water used by man. Nevertheless, the Al contents in water are a subject of growing concern in this country as acidification of the environment tends to lower pH and raise the concentration of Al and other metals in the groundwater. Finnish groundwaters are especially vulnerable to deleterious changes as they are naturally quite acidic and very poorly buffered.

The most harmful gaseous component in bedrock groundwater is radon in a broad zone across southern Finland. As stated above, the distribution of the U content in soil and water is roughly in agreement with that of Rn in indoor air and in water of bedrock wells in the Rn-inflicted areas. In some eastern Finnish communities Rn contents in groundwater may reach a level of 400- 44,400 Bq/L posing a real health hazard. Generally, however, groundwater can only transmit a small proportion of radiation to man (presumably less than 10%), the main part being transmitted through the basements of houses.

## References

Anon, (1979). *Water Quality in Water Utilities in 1977*. Report No. 167, 277 pp. and appendices. National Board of Waters, Finland.
Anon, (1983). *Water Quality in Water Utilities in 1980*. Report No. 226, 255 pp. and appendices. National Board of Waters, Finland.
Anon, (1986). *Water Quality in Water Utilities in 1984*. Report No. 277, 205 pp. and appendices. National Board of Waters, Finland.
Asikainen, M. and Kahlos, H. (1977). *Natural Radioactivity of Ground and Surface Water in Finland*, Raportti STL-A24, 32pp. Institute of Radiation Protection, Helsinki.
Kahma, A. (1973). The main metallogenic features in Finland. *Geol. Survey Finland, Bull.*, **265**, 28.
Lahermo, P. (1984). Hydrogeochemistry in geomedicine. In: J. Låg (ed.), *Geomedical Research in Relation to Geochemical Registrations*, pp.27-45. Universitetsforlaget, Oslo. The Norwegian Academy of Science and Letters.

Lahermo, P. (1987). Atmospheric, geological, marine and anthropogenic effects on groundwater quality in Finland. In: *Proc. Int. Symp. on Groundwater Microbiology: Problems and Biological Treatment*. pp.4-8. IAWPRC, Kuopio, Finland.

Pärkö, A. (1975). Dental caries in the rapakivi granite and olivine diabase areas of Laitila, Finland. *Proc. Finnish Dental Soc.*, 81(1), 55.

Simonen, A. (1980). The Precambrian of Finland. *Geol. Survey Finland, Bull.*, 304, 58.

# 18 Environmental Projects in Czechoslovakia: Influence of Industrial Activities on Plant Production

Vlasta Petríková
*Research Institute for Crop Production, Praha-Ruzyne, Czechoslovakia*

## Summary

*Plants cultivated on ash and mine-dumps do not always have a higher content of heavy metals than those cultivated on soil. Application of slurry is a very efficient treatment in the reclamation of ash and mine-dumps. It is probable that air pollution contaminates food chains more than polluted soils.*

Projects concerning the environment in Czechoslovakia are conducted by the State Commission of Science, Technology and Economics. The relation of the environment to agriculture is included in the projects of the Ministry of Agriculture and the Research Institute for Crop Production.

The negative effects of industrial activity on agriculture are both direct and indirect. The presence of pollutants in the soil is a direct effect preventing rapid reclamation. The most important requirement in the reclamation of industrial waste-sites (ash and mine-dumps) is to increase biological activity. Slurry (from pigs, cattle, poultry) is suitable for this purpose. Large quantities have been used, particularly in ash-dumps. The yields after slurry application were always higher than with mineral fertilisers (Table 1).

The content of heavy metals in crops has also been studied. The results in Table 2 show that plants cultivated on waste-sites (even under extremely bad conditions with no topsoil) do not always have a higher heavy metal content than those cultivated on soil.

The indirect effects of industry on plant production are caused by aerial emissions. These can cause decreased yields and contamination of plants with foreign substances. In Czechoslovakia damage caused to agricultural production from emissions is assessed using measurements of $SO_2$ in air. The quantity of heavy

**Table 1.** *Average yields of some agricultural plants - CU.ha$^{-1}$ (cereal units).*

| Plots | Mine-dumps | | Ash-dumps | |
| | Alfalfa plants | Agricultural | Alfalfa plants | Agricultural |
| --- | --- | --- | --- | --- |
| Control | 26.6 | 18.9 | 9.6 | 8.7 |
| Mineral fertilisers | 31.6 | 30.5 | - | 25.2 |
| Slurry | 38.2 | 37.3 | 36.6 | 42.1 |

**Table 2.** *Average content of heavy metals in some agricultural plants (ppm).*

| | | Plots | |
| Element | Control soil | Mine-dump | Ash-dump |
| --- | --- | --- | --- |
| Cd | 0.185 | 0.140 | 0.110 |
| Hg | 0.020 | 0.040 | 0.029 |
| Pb | 3.740 | 3.330 | 3.900 |
| As | 1.810 | 1.890 | 3.770 |
| Cu | 17.62 | 18.34 | 19.30 |
| Zn | 32.59 | 33.26 | 14.10 |
| Mn | 54.55 | 45.80 | 44.70 |

metal uptake from this source has, as yet, not been determined. It has been found that the content of metals in hay increases, especially during periods without rain. This may be due to the accumulation of dusts which have then not washed away. Indeed, correlations have been shown between the heavy metal content of hay and rainfall; the heavier the rainfall, the lower the content of heavy metals (Cd, Hg, Pb, As) in hay.

# 19 Factors Influencing Selenium Uptake by Some Grass and Clover Species

Louise Arnold, Sheila Van Dorst and Iain Thornton
*Applied Geochemistry Research Group, Geology Department,*
*Imperial College, London SW7 2BP, England*

Selenium uptake by pasture species is dependent on many physical, chemical and biological factors (Levesque, 1974; Smith, 1983; Thornton *et al.*, 1983). These factors include soil pH (Goering *et al.*, 1968), type, the application of fertiliser, and the differing abilities of plant species to accumulate selenium.

Greenhouse pot trials were used to examine the influence of these factors on selenite and selenate accumulation in selected pasture species.

Plants were grown in vermiculite using seven inch diameter flower pots. The pots were treated with 2 x 100 mL applications of sodium selenite or sodium selenate (0.5 µg Se/mL) at pH 4.5 or pH 6.5. Control pots containing no selenium were also prepared. Triplicate pots for each treatment were used. The plants were supplied with Fisons Solinure 5 fertiliser and maintained at 27°C with minimal additional light. Pots were watered as required with tap water adjusted to pH 4.5 or pH 6.5.

Seven pasture species were examined:

| | | | |
|---|---|---|---|
| *Holcus lanatus* | (HL) | *Lolium perenne* | (LP) |
| *Dactylis glomerata* | (DG) | *Trifolium pratense* | (TP) |
| *Trifolium repens* | (TR) | *Festuca ovina* | (FO) |
| *Festuca rubra* | (FR) | | |

Difficulties arise in comparing selenium accumulation data from different plant species as some species grow more rapidly than others. Growth and dilution effects may bias the interpretation of the data. Total selenium accumulated per pot was therefore calculated.

Harvest dry weight varied between species. The largest dry weights were recorded by Holcus lanatus and the smallest by Festuca rubra. Harvest dry weight was in the order:

HL > LP > DG > TP > TR > FO > FR.

Selenium concentrations in the shoots of the seven species were in the orders:

183

**pH 4.5**   Selenite selenium:
FR > TR > TP > LP > FO > HL > DG
Selenate selenium:
TR > TP > DG > FR > LP > FO > HL
**pH 6.5**   Selenite selenium
TR > FR > HL > DG > LP > TP > HL
Selenate selenium
HL > FR > TR > DG > TP > LP > FO

Total selenium levels accumulated per pot by the seven species were in the orders:
**pH 4.5**   Selenite selenium
HL > LP > DG > TP > TR > FO > FR
Selenate selenium
TP > TR > DG > HL > LP > FO > FR
**pH 6.5**   Selenite selenium
HL > TR > DG > LP > TP > FR > FO
Selenate Selenium
HL > DG > TP > LP > TR > FO > FR

## Conclusions

1. The differing ability of plants to accumulate, transform and translocate selenite and selenate may account for the variation in selenium concentration between plant species.
2. At pH 4.5 the grasses accumulate more selenite selenium than selenate (with the exception of Dactylis glomerata), the reverse is found with the two clover varieties.
3. At pH 6.5 all species accumulate greater concentrations of selenite selenium than selenate. In soil the selenite ion forms an insoluble ferric hydroxide-selenite complex which is unavailable for plant uptake. This complex formation does not take place using vermiculite and the selenite ion is available for plant accumulation.
4. Using data from plant uptake experiments the abilities of different plant species to accumulate selenium can be classified and recommendations on sward compositions for low selenium soils can be made.

High levels of sulphate in soils are thought to compete with selenium uptake, especially that of $SeO_4{}^{2-}$, the selenate ion. Selenium uptake may also be affected by applications of nitrogen fertilisers due to higher growth rates.

Pasture species were grown in vermiculite with a nutrient solution and treated with combinations of sulphur and nitrogen containing fertilisers and solutions of selenite and selenate. The fertilisers were applied at rates corresponding to ADAS recommendations and the selenium solutions used were diluted and comparable with concentrations in marginally deficient pasture soils. An Arnold-Hoagland nutrient solution (modified to remove sulphate ions) was used on all pots throughout the experiment as necessary.

Four species of common pasture plants were used:

A. *Trifolium repens* - White Clover
B. *Trifolium pratense* - Red Clover
C. *Dactylis glomerata* - Cocksfoot (Grass)
D. *Lolium perenne* - Perennial Rye (Grass)

Thirty six pots (7 inch) of each species were sown using medium grade vermiculite and watered with tap water (pH 7.5) throughout the experiment.

The first treatment solutions were given as soon as the plants were established and three harvests were taken one month apart. A second treatment was then given and two further harvests were taken of the mature plants. The selenium concentration in the third and fourth harvests before and after the second treatment are shown in Figure 1.

The 12 treatments applied to each species were as follows:

1. Control: nutrient solution (NS) only.
2. Selenite solution ($Na_2SeO_3$) 100 mL of 0.5 ppm      + NS
3. Selenate solution ($Na_2SeO_4$) "     "      + NS
4. Ammonium sulphate solution [$(NH_4)_2SO_4$] 3.2 g in soln.   + NS
5. Ammonium nitrate solution ($NH_4NO_3$) 2.1 g in soln.   + NS
6. Superphosphate fertiliser [$Ca(H_2PO_4)CaSO_4$] 2.0 g in soln.   + NS
7. Selenite solution + Ammonium sulphate solution      + NS
8.   "    "    + Ammonium nitrate solution      + NS
9.   "    "    + Superphosphate fertiliser      + NS
10. Selenate solution + Ammonium sulphate solution      + NS
11.   "    "    + Ammonium nitrate solution      + NS
12.   "    "    + Superphosphate fertiliser      + NS

Each treatment was repeated in triplicate on separate pots.

## Conclusions

1. Selenite accumulation was greater than that of selenate in all species and treatments.
2. The grass species showed slightly higher accumulations of selenium than the clover species.
3. There was no noticeable effect of sulphate suppressing selenate uptake at these low selenium levels. There was some evidence of sulphate suppressing selenite uptake in grasses.
4. There was a marked increase in the growth rate of grasses receiving nitrogen fertilisers. Clovers receiving nitrogen did not show this increase in growth and this is strong evidence for root nodulation and active nitrogen fixation even though the plants were grown in vermiculite.
5. Clover species showed little or no reduction in selenium concentration with increased growth due to fertiliser application. Dactylis glomerata showed little variation in selenium concentration with fertiliser treatments. However due to the increased growth of the nitrogen-treated plants, the total uptake of selenium in the

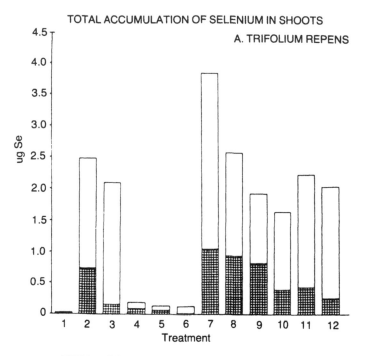

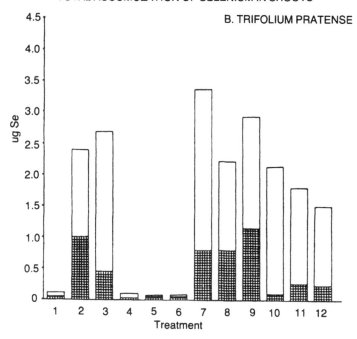

**Figure 1.** *Selenium concentrations in test plants at various stages of the experiment.*

# TOTAL ACCUMULATION OF SELENIUM IN SHOOTS

C. DACTYLIS GLOMERATA

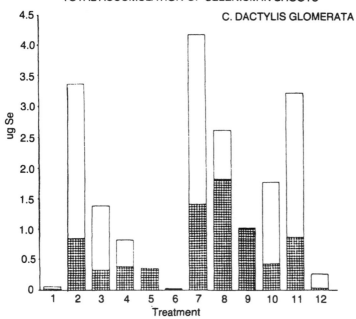

# TOTAL ACCUMULATION OF SELENIUM IN SHOOTS

D. LOLIUM PERENNE

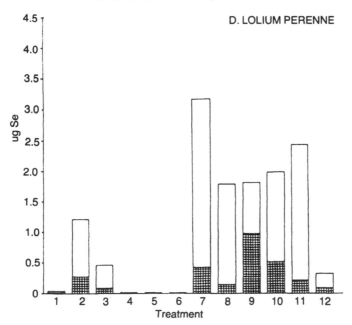

shoots was greatly increased. Only Lolium perenne showed a strong reduction in selenium concentration due to increased growth in the selenite and fertiliser treatments. Large applications of nitrogen fertilisers may lower the concentration of selenium in herbage quite markedly, although the overall uptake may not be affected.

## References

Geering, H.R., Cary, E.E., Jones, L.H.P. and Allaway, W.H. (1968). *Soil Sci. Am. Proc.*, **32**, 35-40.

Levesque, M. (1974). *Can. J. Soil Sci.*, **54**, 63-68.

Smith, C.A. (1983). Ph.D. Thesis, University of London.

Thornton, I., Kinniburgh, D.G., Pullen, G. and Smith, C.A. (1983). In: D.D. Hemphill (ed.), *Trace Substances in Environmental Health XVII, 1983*. University of Missouri, Columbia, USA.

# 20 Aluminum Intake and Its Effects

Seymour G. Epstein
*The Aluminum Association, Inc., 900 19th St., NW, Washington, DC 20006, USA*

## Summary

*The Aluminum Association has followed closely the developing knowledge of health effects of aluminum and aluminum compounds, particularly the recent neurological implications. Through our research efforts, through continuing surveillance of the world's scientific literature, and through discussions with leading researchers in the field, we draw the following conclusions regarding the relationship between aluminum and Alzheimer's disease: (1) the cause (or causes) of Alzheimer's disease is not known, (2) the biological significance of aluminum in the brain is not understood, (3) aluminum is poorly absorbed by the body, and (4) ordinary environmental exposure to aluminum is safe.*

## Introduction

In 1886, the first practical and economic process for producing aluminum metal was discovered. Cookware was the first commercial application for the new metal. Shortly thereafter, claims of various adverse health effects from exposure to aluminum began to appear, but they did not originate from nor were they supported by scientific literature.

The first comprehensive treatise on aluminum compounds in food was published in 1928 (Smith, 1928). The author presented considerable evidence that aluminum is not injurious to health, but added, "Unfortunately, this question has become controversial by reason of conflicting commercial interests".

Starting in 1955, under sponsorship by the Aluminum Association, research teams at the Kettering Laboratory of the University of Cincinnati have periodically searched out and reviewed the world's literature on aluminum and health and published their findings. The first conclusion drawn (Campbell *et al.*, 1957) was that there is no need for concern among the public regarding hazards to human health from exposure to aluminum products. The review was updated in 1974 and the basic conclusions were reaffirmed (Sorenson *et al.*, 1974).

Since 1980, the literature reviews at the University of Cincinnati have been continuous. Recent reports (Krueger et al., 1984; Krueger and Clark, 1987) raise questions about long-time exposure to medicinal doses of aluminum compounds, and stress the importance of the ligands with which aluminum is associated since these affect the bioavailability and distribution of aluminum in the body.

When allegations concerning neurological effects of aluminum began appearing about 10 years ago the Association again turned to Kettering Laboratory, this time for an in-depth review of the literature on the neurological implications of aluminum. In addition, discussions began with leading investigators.

The Kettering report (Cooper et al., 1981) concluded that there was no direct clinical or experimental evidence that aluminum is neurotoxic to humans or animals under ordinary conditions of environmental exposure. It was also noted that gaps exist in the knowledge of the significance of aluminum in the human body. This was thought to be principally because aluminum was not generally regarded as posing a health problem and, hence, drew little scientific interest or study.

The need was recognised for basic information on the way aluminum gets into the body, how much typically is absorbed, where it goes, and what effects it may produce. The Association set into place a long-range research program to provide information on these subjects. In addition, literature surveillance and contacts with investigators have been continued.


## Aluminum Intake

Aluminum is the third most abundant chemical element in nature, constituting nearly 8% of the earth's crust. Its various forms and their prevalence have been reviewed previously (Epstein, 1985). It is omnipresent; food, water and air all contain measurable amounts of the element. Thus it is virtually impossible to avoid exposure to aluminum. However, the human body appears to have adjusted well to its everyday exposures.

Table 1 lists the sources of aluminum which might be normally ingested. Estimates are given for the daily amounts of aluminum typically expected from those sources where data are available. The adult body content has been estimated at 295 mg (Skalsky and Carchman, 1983).

*Food*
The Food and Drug Administration (FDA) publishes a *Total Diet Study* which lists 235 foods commonly eaten in the US. Using the latest list (Pennington, 1983), the aluminum content of typical American diets was estimated to be from 3 to over 100 mg of aluminum daily, with most adults probably consuming between 20 to 40 mg/day (Greger, 1985). Amounts from earlier studies were estimated in the range of 10 to 100 mg/day. Researchers at FDA have recently analysed the aluminum content of an actual total diet collection and the results should be published shortly.

Aluminum in the diet comes from several sources:
*Natural Content* - Because of the omnipresence of aluminum in soils and waters, virtually all foods contain measurable amounts of natural aluminum. The average

**Table 1.** *Estimated daily intake of aluminum from various sources.*

| Category | Source | mg Al/day |
|---|---|---|
| Food | Natural Content | 2 - 10 |
| | Intentional Additives (FDA-approved Al compounds) | 20 - 50+ |
| | Unintentional Additives (from metallic Al products) | 3.5[*] |
| | Total Diet | 3 - 100+ |
| Water | Natural Content, Alum | <1 |
| Air | Dust, Smoke, Toiletries, Sprays | <1 |
| Drugs | Antacids | 50 - 1,000+ |
| | Buffered Aspirin | 10 - 100+ |

[*] Maximum value under "worst-case scenario".

adult American probably takes in between 2 to 10 mg of Al/day as "natural" content of the foods eaten.

*Intentional Additives* - A much larger amount of aluminum is typically ingested in the form of intentional additives. These are FDA-approved aluminum compounds used as preservatives, colouring agents, leavening agents, *etc.* in a wide variety of foods. The amount of aluminum ingested in these compounds typically ranges from about 20 to over 50 mg, but can vary widely depending on the dietary preferences.

*Unintentional Additives* - The third source of aluminum in food comes from unintentional additives, *i.e.* that which is added to the food through cooking, packaging and handling in metallic aluminum products. Based on a comprehensive study involving a variety of foods and analyses by state-of-the-art atomic absorption techniques (Greger *et al.*, 1985), Greger (1985) estimated that about 3.5 mg Al/day could be added to the diet from this source if *all* cooking and handling is done with uncoated aluminum products. This represents only a small percentage of the average dietary intake of aluminum, even under conditions of highest use.

### Water

Virtually all water contains small amounts of aluminum. Most aluminum compounds are relatively insoluble in neutral water, and there is usually less than 1 mg/L of aluminum in these waters (Miller *et al.*, 1984). At this level, a typical intake of 1 to 2 L of water daily will contribute less than 1 mg of aluminum to the diet.

### Air

Aluminum compounds are present in air coming from various sources, such as dry soil, coal combustion and cigarette smoke. They are in household and workplace environments, particularly as aluminum oxide dust. Although amounts present vary from location to location, they are relatively small. The American Conference of Governmental Industrial Hygienists recommends a threshold limit value for aluminum and aluminum oxide in air of 10 mg/m$^3$, the same level as for other nuisance dusts. There are few environments where this level is exceeded. At ambient levels of aluminum in air, typically a few μg/m$^3$, intake from this source is far less than 1 mg/day based on inhalation of about 20 m$^3$/day of air.

*Drugs*

Most antacids contain aluminum compounds, and one antacid tablet may contain 50 mg or more of aluminum. It is not unusual for a person with a stomach disorder to consume more than 1,000 mg of aluminum per day. Those with chronic stomach conditions, such as peptic ulcers, may take multiples of such doses over a considerable length of time. A buffered aspirin tablet may contain about 10 to 20 mg of aluminum. Aluminum compounds are also in many vaccines as adjuvants. Indeed, the World Health Organisation and the Food and Drug Administration recognise aluminum's value as an adjuvant for injectable drugs, and define permissible levels with full knowledge that the aluminum is going directly into the blood stream. Clearly, the amount of aluminum taken into the body through drugs can dwarf the amount from normal ingestion and inhalation.

The gastrointestinal tract is only very slightly permeable to aluminum and provides a relatively effective barrier to its absorption (Skalsky and Carchman, 1983). Very little is known about human absorption and metabolism of aluminum. Such studies have been hampered by the lack of a useful aluminum radioisotope for tracer studies and the difficulties in determining the trace levels of chemical aluminum present in the blood and body tissues (Savory and Wills, 1983). In particular, contamination from the ever-present aluminum in the environment makes analysis even more difficult and can lead to falsely elevated measurements (Brown *et al.*, 1983).

In a recent review an attempt was made to compensate for the lack of knowledge regarding aluminum in the body by substituting data for other, but chemically similar, metallic ions (Ganrot, 1986). By these comparisons, a more detailed picture of the behaviour of aluminum in the body was simulated.

Homeostatic processes strive to maintain an equilibrium between the body and its environment. Most, if not all, of the aluminum that is ingested from whatever source is excreted. This was demonstrated in a study at the University of Wisconsin (Greger and Baier, 1983) in which eight adult males were fed 125 mg of aluminum per day by supplementing their controlled dietary intake with soluble aluminum lactate. On average, no retention of aluminum was detected. Earlier studies had shown some retention of aluminum at daily intakes exceeding 1,000 mg/day (Skalsky and Carchman, 1983). Thus, apparently at some level of exposure between 125 and 1,000 mg/day, the body may absorb aluminum faster than it can excrete it and aluminum may begin to accumulate. A comparison of intake levels and body balance is shown in Figure 1.

192

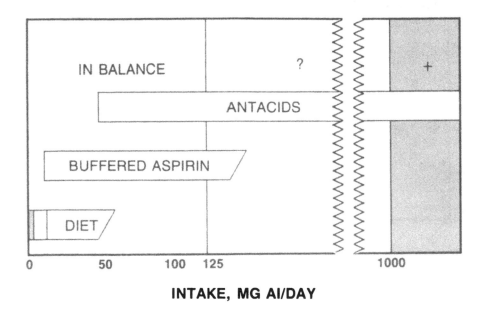

**INTAKE, MG Al/DAY**

**Figure 1.** *Typical aluminum intakes and body balance.*

### Effects

Aluminum is present in the body - in all organs, tissues and fluids - from birth. Because of this some investigators have suspected that aluminum may be an essential element but have been unable to produce the aluminum-free environment needed to substantiate this supposition. This has also made it difficult to ascertain the biological significance of aluminum in the body.

Figure 2 graphically depicts the distribution of aluminum among the organs of the human body (Skalsky and Carchman, 1983). The graph is constructed as the logarithm of the ratio of the percent aluminum in the organ to the organ weight as a percent of the body weight, which has the effect of "normalising" the values. The majority of the organs and tissues are shown to contain aluminum in relative proportion to their masses. Blood, brain and adipose contain significantly less aluminum than would be expected based on tissue mass, whereas bone and lung tissue contain more aluminum than expected based on their relative masses. This can be interpreted as an indication that aluminum does not accumulate in blood, brain and adipose to any great extent but does accumulate in bone and in the lung, with the data suggesting that bone may offer a major reservoir for aluminum absorbed into the body.

As discussed previously (Epstein, 1985) the only generally recognised health effect of aluminum intake by humans with normal kidney function is phosphate depletion. The remainder of this paper will concentrate on the present controversies

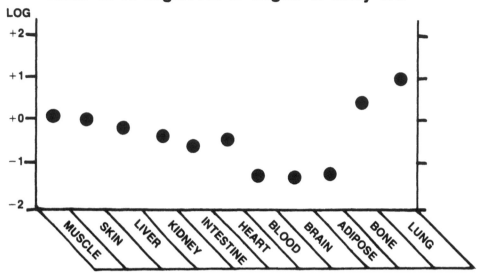

**Figure 2.** *Aluminum distribution in body organs.*

regarding a possible relationship between aluminum in the body and Alzheimer's disease.

The principal laboratory findings linking aluminum with Alzheimer's disease were that (1) injections of aluminum salts directly into the brains of animals produced neurofibrillary tangles and (2) the brains of human victims of Alzheimer's disease contained more aluminum than did "normal" brains.

*The implications of aluminum - animal studies*

Klatzo *et al.* (1965) reported neurofibrillary changes in animal brains resulting from direct injection of aluminum salts. Using electron microscopy, investigators subsequently found that the filaments artificially produced in animals were *single* strands and were far different from the *paired helical filaments* (PHF) observed in human brains (Wisnieski *et al.*, 1970, 1976).

Wisnieski *et al.* (1980) later verified that "aluminum-induced" tangles in animals are chemically, as well as structurally, different from the paired helical filament tangles found in humans. In addition, the aluminum injections did not produce senile plaques and both PHF and plaques must be observed for a diagnosis of Alzheimer's disease to be seriously considered.

Investigators continue to study the effects of direct injections of aluminum powder and aluminum salts into animal brains in order to bypass the body's controls. However, while experimentally convenient, the situation created is most artificial and the observed results have questionable relevance to aluminum *per se*. Irby (1984) found that aluminum powder added to a saline solution simulating brain fluid produced large amounts of hydrogen gas and changes in solution pH. What effects similar evolution of hydrogen would have in animal brains is speculative.

At forums where these issues have been discussed the consensus view is that there is no valid animal model for Alzheimer's disease.

*The implications of aluminum - human studies*

Using atomic absorption spectroscopy, Crapper (McLachlan) *et al.* (1973, 1976, 1978) measured the aluminum content of a number of samples of human brains, both "normal" and demented. The "normal" content for adults averaged about 2 μg Al/g of dry tissue. Alzheimer's brains appeared to have about twice that amount on average, with considerable scatter and overlap in the data bands.

Following these reports, several studies were conducted attempting to verify the observations. These yielded conflicting results which show aluminum present in all brains, from birth, but do not support a direct correlation with disease state (Trapp *et al.*, 1978; McDermott *et al.*, 1979; Traub *et al.*, 1981; Markesbery *et al.*, 1981).

A series of studies have since been conducted in which 18 elements, including aluminum, were measured simultaneously by neutron activation analysis of brain samples from 12 victims of Alzheimer's disease and 28 control (non-demented) brains. The findings were that aluminum increases slightly in the brain with age, but not with the disease (Markesbery *et al.*, 1983). Three other elements - bromine, cobalt and mercury - were significantly elevated in the diseased brains (Ehmann *et al.*, 1982). In a subsequent paper it was reported that eight elements, but not aluminum, were significantly different in Alzheimer's brains compared to the control brains (Ehmann *et al.*, 1986).

Hershey *et al.* (1983) analysed the cerebrospinal fluid from a number of patients, including those with Alzheimer's disease. No elevation of aluminum was found, but increased levels of silicon were detected in a significant number of the Alzheimer's patients.

Shore and Wyatt (1983) found no elevations of aluminum concentrations in the serum, cerebrospinal fluid, or hair of Alzheimer's patients and suggested that aluminum alone is not a causal agent but rather a marker of degenerating neurons. The authors also knew of no reports of beneficial effects of aluminum-restricted diets on the course of Alzheimer's disease.

If indeed brain aluminum were elevated in Alzheimer's victims, one still would not know whether the element was contributing to the disease or whether its presence was a result of the disease. Other diseases characterised by cell damage are known to attract elements, such as copper and calcium, to disease sites. Furthermore, many renal failure patients exhibit high levels of tissue and brain aluminum from the therapeutic treatments, but do not develop Alzheimer's disease or the Alzheimer-like pathological changes (Alfrey *et al.*, 1976).

Among scientists at the forefront of research into Alzheimer's disease, interest has apparently waned as to whether or not aluminum is elevated in the brains of Alzheimer's victims. The finding of Crapper McLachlan remains controversial and unreplicable outside his laboratory. There now appears to be greater acceptance of the premise that aluminum does slightly accumulate in the brain as part of the natural aging process (Ganrot, 1986). For the past several years considerably more attention has been given to localisation of aluminum in the brain, *i.e.* where it appears at the cellular level.

Perl and Brody (1980) found that aluminum and other elements in disease-damaged cells were present at levels somewhat above background in samples of three Alzheimer's brains. However, the technique was not quantitative and the amounts of the elements present could not be determined.

From a collaborative effort involving two hospitals in Newcastle upon Tyne and the University of Cambridge, Candy *et al.* (1986) reported finding an aluminum-silicon compound in the cores of senile plaques found in both Alzheimer and non-Alzheimer patients. Others (Stern *et al.*, 1986; Wisnieski *et al.*, 1986) were unable to reproduce this finding even with the same tissue samples. However, the implications of this possibility are now being discussed widely.

## Discussion

During the past few years several excellent reviews on possible causes of Alzheimer's disese have been published in scientific and medical literature. Symposia have been held, and more are planned, on the possible role of environmental aluminum in this disease.

A conference on Aluminum Analysis in Biological Materials, co- sponsored by the Aluminum Association and the University of Virginia, was held in June 1983. Researchers in attendance agreed on two points: (1) the cause (or causes) of Alzheimer's disease is not known and (2) the biological significance of aluminum in the brain is not understood.

In a review article on Alzheimer's disease, Wurtman (1985) wrote "No-one knows its cause or how to stay its inexorable course." Six models which now underlie most research on Alzheimer's disease were presented and discussed: (1) faulty genes, (2) abnormal proteins, (3) an infectious agent, (4) an environmental toxin, (5) inadequate blood flow and (6) a neurotransmitter deficiency.

The search for the cause of Alzheimer's disease was further discussed by Katzman (1986). Regarding aluminum, Katzman concluded, "There is no evidence that exposure to such sources of exogenous aluminum as aluminum antacids, antiperspirants, or even the large amount used in renal dialysis increases the risks of Alzheimer's disease. Thus, a direct relation between exogenous aluminum and aluminum deposits in the brains of persons with Alzheimer's disease has not been established."

In September 1986 the National Institute on Aging (NIA) and the American Association of Retired Persons (AARP) jointly sponsored a research conference on Trace Metals, Aging and Alzheimer's Disease. Khachaturian (1986), the conference organiser, later wrote, "At the conference there was a general consensus on the following four points: (a) the nature of the materials in the classic morphological lesions of Alzheimer's disease are becoming increasingly clear; (b) there is clear evidence that aluminum can be a neurotoxin; (c) there is agreement that aluminum does accumulate in at least specific regions of Alzheimer's disease brains and that the chemistry of aluminum is so complex as to make the interpretation of data difficult just from the molecular level, and (d) there is no clear evidence that the removal of aluminum or the prevention of its accumulation would alter the course of Alzheimer's disease, which is the question with which we ultimately have to wrestle." Proceedings of the conference are expected later this year.

The November/December 1986 issue of *Neurobiology of Aging* was a special issue devoted to "Controversial Topics on Alzheimer's Disease: Intersecting Crossroads". In one or two review articles on aluminum Crapper McLachlan (1986) concluded that the evidence does not support an etiological role for aluminum in Alzheimer's disease. He cited two opposing points of view on the functional significance of aluminum in Alzheimer's disease: (1) aluminum merely accumulates passively in neurons compromised by the Alzheimer degenerative process and the accumulation is of no significance to the mechanisms of the disease; or alternatively, (2) aluminum is a plausible candidate for a neurotoxic environmental factor acting in the pathogenesis of the degenerative processes.

The commentaries on the review ranged widely. One paper (Wisnieski *et al.*, 1986) maintained there is no evidence for aluminum in the etiology and pathogenesis of Alzheimer's disease.

The second review involving aluminum presented a novel hypothesis that Alzheimer's disease may begin in the nose and may be caused by aluminosilicates (Roberts, 1986). However, Terry (1986), commenting on the hypothesis, wrote that it stands on two fragile legs.

Considerable attention is now being given to new findings that strengthen the evidence of a genetic role in Alzheimer's disease and a demonstrated link with Down's Syndrome (Grundke-Iqbal, *et al.*, 1986; Kosik *et al.*, 1986; Robakis *et al.*, 1986; Robakis *et al.*, 1987). However, in a recent paper on etiology of the disease, the authors contend that present evidence is sufficiently impressive to justify the provisional conclusion that a virus is responsible for Alzheimer's disease (Mozar, Dileep and Howard, 1987). In either case the door is kept open that environmental factors, such as aluminum, may play a role in the development of the disease.

One final comment: in the last few years, three case-control studies were performed to assess the possible roles of various factors in the development of Alzheimer's disease (Heymann *et al.*, 1984; French *et al.*, 1985; Amaducci *et al.*, 1986). Included in the studies were occupational exposures and exposures to aluminum-containing antacids. There were no indications that such exposures increased the risk of the disease in the populations studied.

## Conclusions

Through research efforts, through continuing surveillance of the world's scientific literature, and through personal contacts and discussions with leading researchers in the field, several conclusions have been drawn concerning the present state of knowledge regarding the relationship between aluminum and Alzheimer's disease (Epstein, 1985). These, listed below, continue to appear valid.
- The cause (or causes) of Alzheimer's disease is not known.
- The biological significance of aluminum in the brain is not understood.

- Aluminum is poorly absorbed by the blood.
- Ordinary environmental exposure to aluminum is safe.

The Aluminum Association continues to fund research into the role of aluminum, if any, in the human body. The scientific and medical literature on the health effects of aluminum is continually reviewed and discussions are held periodically with investigators at universities and research institutions to remain at the forefront of this subject. A comprehensive monograph on all aspects of aluminum and health is now being prepared, and should be published early next year.

## References

Alfrey, A.C., LeGendre, G.R. and Kaehny, W.D. (1976). The dialysis encephalopathy syndrome. *New England J. Medicine*, **294**, 184-188.

Amaducci, L.A. *et al.*. (1986). Risk factors for clinically diagnosed Alzheimer's disease. *Neurology*, **36**, 922-931.

Brown, S. *et al.* (1983). Specimen collection - sources of contamination. *Proceedings of Conference on Aluminum Analysis in Biological Materials*, pp.60-66. Charlottesville, VA.

Campbell, I.R. *et al.* (1957). Aluminum in the environment of man. *A.M.A. Arch. Industrial Health*, **15**, 359-448.

Candy, J.M. *et al.* (February 15, 1986). Aluminosilicates and senile plaque formation in Alzheimer's disease. *The Lancet*, February 15, 354-356.

Cooper, G.P., Krueger, G.L. and Widner, E.M. (1981). Neurotoxicity of Aluminum. Final Report to the Aluminum Association, May 22.

Crapper (McLachlan), D.R., Krishnan, S.S. and Dalton, A.J. (1973). Brain aluminum distribution in Alzheimer's disease and experimental neurofibrillary degeneration. *Science*, **180**, 511-513.

Crapper (McLachlan), D.R., Krishnan, S.S. and Quittkat, S. (1976). Aluminum, neurofibrillary degeneration and Alzheimer's disease. *Brain*, **99**, 67-80.

Crapper (McLachlan), D.R., Karlik, S. and DeBoni, V. (1978). Aluminum and other metals in senile (Alzheimer) dementia. *Aging*, **7**, 471-485.

Crapper (McLachlan), D.R. (1986). Aluminum and Alzheimer's disease. *Neurobiology of Aging*, **7**, 525-532.

Ehmann, W.D. and Markesbery, W.R. *et al.* (1982). Trace elements in human brain tissue by INAA. *J. Radioanalytical Chem.*, **70**, 57- 65.

Ehmann, W.D. and Markesbery, W.R. *et al.* (1986). Brain trace elements in Alzheimer's disease. *Neurotoxicology*, **7**, 197-206.

Epstein, S.G. (1985). Aluminum in nature, in the body and its relationship to human health. *Trace Elements in Medicine*, **2**, 14- 18.

French, L.R. *et al.* (1985). A case-control study of dementia of the Alzheimer type. *Am. J. Epidemiology*, **121**, 414-420.

Ganrot, P.O. (1986). Metabolism and possible health effects of aluminum. *Environ. Health Persp.*, **65**, 363-441.

Greger, J.L. (1985). Aluminum content of the American diet. *Food Tech.*, **39**, 73-80.

Greger, J.L., Goetz, W. and Sullivan, D. (1985). Aluminum levels in foods cooked and stored in aluminum pans, trays and foil. *J. Food Protection*, **48**, 772-777.

Greger, J.L. and Baier, M.J. (1983). Excretion and retention of low or moderate levels of aluminum by human subjects. *Food Chemical Toxicol.*, **21**, 473-477.

Grundke-Iqbal, I., Iqbal, K., Quinlan, M., Tung, Y.C., Zaidi, M.S. and Wisnieski, H.M. (1986). Microtuble-associated protein tau: a component of Alzheimer paired helical filaments. *J. Biol. Chem.*, **261**, 6084-6089.

Hershey, C.O. *et al.* (1983). Cerebrospinal fluid trace element content in dementia, clinical, radiologic, and pathologic correlations. *Neurology*, **33**, 1350-1353.

Heyman, A. *et al.* (1984). Alzheimer's disease: a study of epidemiological aspects. *Ann. Neurology*, **15**, 335-341.

Irby, E.C. (1984). Unpublished data. Reynolds Metals Company.

Katzman, R. (1986). Alzheimer's disease. *New England J. Medicine*, **314**, 964-973.

Khachaturian, Z.S. (1986). Aluminum toxicity among other views on the etiology of Alzheimer disease. *Neurobiology of Aging*, **7**, 537- 539.

Klatzo, I., Wisnieski, H.M., Stricher, E. (1965). Experimental production of neurofibrillary degeneration. I. Light microscopic observations. *J. Neuropathology Exp. Neurology*, **24**, 187-199.

Kosik, K.S., Joachim, C.L. and Selkoe, D.J. (1986). The microtuble associated protein, tau, is a major antigenic component of paired helical filaments in Alzheimer's disease. *Proc. National Acad. Sciences USA*, **83**, 4044-4048.

Krueger, G.L. *et al.* (1984). The health effects of aluminum compounds in mammals. *CRC Critical Reviews in Toxicology*, **13**, 1- 24.

Krueger, G.L., Clark, R.A., Widner, E.M. (1987). Effects of aluminum-containing compounds in mammals: a critical review of the literature published in 1986. Final Report to the Aluminum Association.

Markesbery, W.R and Ehmann, W.D. *et al.* (1981). Instrumental neutron activation analysis of brain aluminum. *Ann. Neurology*, **10**, 511-516.

Markesbery, W.R., Ehmann, W.D. and Alauddin, M. (1983). Bulk brain aluminum concentrations in Alzheimer's disease. *Proc. Conference on Aluminum Analysis in Biological Materials*, pp.103- 110. Charlottesville, VA.

McDermott, J.R. *et al.* (1979). Brain aluminum in aging and Alzheimer's disease. *Neurology*, **29**, 809-814.

Miller, R.G. *et al.* (1984). The occurrence of aluminum in drinking water. *J. Am. Water Works Assoc.*, January, 84-91.

Mozar, H.N., Dileep, G.B. and Howard, J.T. (1987). Perspectives on the etiology of Alzheimer's disease. *J. Am. Medical Assoc.*, **257**, 1503-1507.

Pennington, J.A.T. (1983). Revision to the total diet study. *J. Am. Dietary Assoc.*, **82**, 166-172.

Perl, D.P. and Brody, A.R. (1980). Alzheimer's disease: x-ray spectrometric evidence of aluminum accumulation in neurofibrillary tangle-bearing neurons. *Science*, **208**, 297-299.

Perl, D.P. (1985). Relationship of aluminum to Alzheimer's disease. *Environ. Health Persp.*, **63**, 149-153.

Robakis, N.K., Wolfe, G., Ramakrisha, N. and Wisnieski, H.M. (1986). Isolation of a cDNA clone encoding the Alzheimer's disease and Down syndrome anyloid peptide. *Neurochemistry of Aging*. Banbury Conference Proceedings, (in press).

Robakis, N.K., Wisnieski, H.M. *et al.* (1987). Chromosome 21q21 sublocalisation of the gene encoding the beta-amyloid peptide present in cerebral vessels and neuritic (senile) plaques of people with Alzheimer's disease and Down syndrome. *Lancet*, **1**, 384-385.

Roberts, E. (1986). Alzheimer's disease may begin in the nose and may be caused by aluminosilicates. *Neurobiology of Aging*, **7**, 561- 567.

Savory, J. and Wills, M.R. (1983). Analytical techniques for the measurement of aluminum in biological materials. *Proc. Conference on Aluminum Analysis in Biological Materials*, pp.1-14. Charlottesville, VA.

Shore, D. and Wyatt, R.J. (1983). Aluminum and Alzheimer's disease. *J. Nervous Mental Disease*, **171**, 553-558.

Skalsky, H.L. and Carchman, R.A. (1983). Aluminum homeostasis in man. *J. Am. Coll. Toxicology*, **2**, 405-423.

Smith, E.E. (1928). *Aluminum Compounds in Foods*. Paul B. Hoeber Inc., New York.

Sorenson, J.R. *et al.* (1974). Aluminum in the environment and human health. *Environ. Health Persp.*, **8**, 3-95.

Stern, A.J., Perl, D.P. *et al.* (1986). Investigation of silicon and aluminum content in isolated senile plaque cores by laser microprobe mass analysis (LAMMA). *J. Neuropathology Experimental Neurology*, **45**, 361.

Terry, R.D. (1986). Does Alzheimer's disease spread, and is it causally related to aluminum? *Neurobiology of Aging*, **7**, 570.

Trapp, G.A. *et al.* (1978). Aluminum levels in brain in Alzheimer's disease. *Biological Psychiatry*, **13**, 709-718.

Traub, R.D. *et al.* (1981). Brain destruction alone does not elevate brain aluminum. *Neurology*, **31**, 986-990.

Wisnieski, H.M., Moretz, R.C. and Iqbal, K. (1986). No evidence for aluminum in etiology and pathogenesis of Alzheimer's disease. *Neurobiology of Aging*, **7**, 532-535.

Wisnieski, H.M., Terry, R.D. and Hirano, A. (1970). Neurofibrillary pathology. *J. Neuropathology Experimental Neurology*, **29**, 163-176.

Wisnieski, H.M., Narang, H.K. and Terry, R.D. (1976). Neurofibrillary tangles of paired helical filaments. *J. Neurological Science*, **27**, 173-181.

Wisnieski, H.M., McDermott, J.R. and Iqbal, K. (1980). Aluminum- induced neurofibrillary changes: its relationship to senile dementia of the Alzheimer's type. *Aluminum Neurotoxicity*, pp.121- 124. Pathotox Publishers, Inc., Park Forest South, Illinois.

Wurtman, R.J. (1985). Alzheimer's disease, *Scientific American*, 62-74.

# 21 A European Overview of Aluminium in Drinking Water

A.M. Simpson,
*Laporte Inorganics, PO Box 2, Moorfield Road,*
*Widnes, Cheshire WA8 0JU, England*
*and*
C.J. Sollars and R. Perry,
*Public Health Engineering, Imperial College, London SW7 2BU, England*

## Introduction

As the third most common element, constituting 8% of the Earth's crust, aluminium is found universally in soils, waters and air, and is continually moving through these media due to the action of natural forces. Ultimately, both wind-borne and some water- borne aluminium will be deposited on or be taken up by vegetation. Where plants eaten by food animals or directly by man are concerned, the aluminium content of these will eventually find its way into humans (Dinman, 1984). In addition, consumption of drinking water directly or indirectly through food preparation will lead to ingestion of the natural aluminium content. Where drinking water supplies are treated with aluminium sulphate, this may lead in some locations to an increase in potable water aluminium concentrations above those found in the raw water.

Increasing public and medical interest has recently arisen concerning a possible association between human exposure to aluminium from various sources, such as drinking water, and the incidence of Alzheimer's disease in developed countries (Epstein, 1985; Krishnan and Crapper McLachan, 1985; Edwardson *et al.*, 1986). At present, research on Alzheimer's disease is still being actively pursued in several countries. Lewis and Waddington (1986), recently reviewing the current situation, concluded that the link, if any, between Alzheimer's disease and aluminium in drinking water was a subject of considerable uncertainty, but pointed out that this was of great importance to the water industry and that further research was clearly needed. Indeed, WHO is already encouraging the development of relevant epidemiological investigations (*ibid.*).

*Aluminium sulphate in water treatment*

The application of alum, the forerunner of aluminium sulphate, for the treatment of water, mordanting of dyes and the fireproofing of materials has been known to man for 4,000 years. Since the mid 1800's aluminium sulphate has been manufactured by reacting sulphuric acid with various forms of bauxite, and has become the conventional coagulant for water treatment across the world. As demand for potable water increases year by year, so an increasing proportion is drawn from surface sources, which are becoming fully utilised. Water from surface sources frequently contains undesirable constituents in the form of material in suspension, colour, coliform organisms, *etc*. In order to render the water acceptable for domestic or industrial uses it is necessary to remove these contaminants by one or more of the following processes: storage, filtration without chemical treatment, clarification by coagulation and sterilisation.

Storage of water in reservoirs allows for stabilisation with some natural settlement of the suspended matter, and enables natural oxidation processes to reduce the concentration of pathogenic bacteria. Nevertheless, storage provides at best only a partial and limited degree of purification, and is in practice followed by filtration of the settled water by biological or mechanical means. Slow sand filtration provides an excellent means of purification but suffers from disadvantages such as low filtration rates and extensive and costly space and labour requirements.

A widely employed alternative is water clarification followed by filtration. An essential step in this process is the addition of chemical coagulants which react with natural or added alkalinity in the water to produce a gelatinous precipitate.

Aluminium sulphate is widely used to provide the gelatinous precipitate of hydroxide which by physical and electrochemical means entrains the suspended and colloidal matter present in the water. Although a number of other coagulants are in use, for many locations the use of aluminium sulphate or other aluminium-based salts is indispensible to the production of drinking water to a standard demanded by today's consumers.

Within this perspective, the authors have undertaken an extensive survey of prevailing concentrations of aluminium in drinking water supplies in a number of European countries. The objectives of the survey were:

- to assess the background situation, including current water quality standards with respect to aluminium and aluminium content of natural waters;
- to collate and evaluate drinking water aluminium data from Western Europe and the USA, with particular emphasis on the relationship between use of aluminium-based coagulants during treatment and residual aluminium concentrations;
- to gain an overview of the current research position on aluminium in drinking water.
- to compare human aluminium intake from drinking water with that from other sources.

202

## Aluminium in Raw and Treated Waters

### Water quality standards

In order to provide a basis for evaluating the aluminium data for raw and treated waters presented in this paper, Table 1 details water quality standards with respect to aluminium which are in force in the countries surveyed. To the authors' knowledge, no standards have yet been promulgated in the countries involved concerning aluminium concentrations in surface or groundwaters used for potable supply abstraction, so Table 1 refers exclusively to drinking water standards.

The table reveals considerable uniformity among the standards given, and for EEC countries this clearly reflects the influence of the 1980 Drinking Water Directive (Council of the European Communities, 1980) on the legislation of member states and other neighbouring countries. The lack of a WHO MAC value, and the absence of a legal limit for the USA is perhaps indicative, however, of the fact that until recently aluminium in potable water was not generally regarded as a matter of serious concern in health terms, but this situation may change if a clear link between aluminium in potable water and Alzheimer's disease is established.

Table 1 will be referred to throughout this paper as appropriate. It should also be noted that throughout this report, the word "alum" is used to refer to aluminium sulphate, as that term is conventionally understood in water treatment terms.

### Aluminium content of natural waters

No comprehensive surveys of aluminium concentrations in rainwater have been previously reported, but values from recent literature suggest that concentrations are generally less than 0.05 mg/L, except where precipitation quality is affected by industrial emission (Cawse, 1974; Talbot and Elzerman, 1985).

Aluminium concentrations in river water can vary widely, and a range of 12-2,250 µg/L has been reported from North American rivers (Durum and Haffty, 1963). In areas where inputs of acidic deposition have occurred, surface waters with low average buffering capacity have exhibited depressed pH values, and consequent reductions in fish stocks have been attributed to elevated aluminium concentrations in the waters concerned (Odonnell et al, 1984). In such areas, many lakes and streams exhibit aluminium concentration with the 100-800 µg/L range (Driscoll and Newton, 1985; Miller and Andelman, 1986)

The aluminium content of groundwaters received very little study prior to the recognition of the problems of acidification, and only in the last few years have the effects od acidification on groundwater trace metal content begun to attract attention, especially in Sweden and Norway. A major concern motivating much of this research is that groundwaters are widely used for small- scale potable water supplies in many of the sparsely populated areas affected (Hultberg and Johansson, 1981; Dickson, 1986; Henriksen and Kirkhusimo, 1986; Taylor et al., 1986). Many of the concentrations reported in these studies are clearly in excess of the EEC maximum acceptable concentration (MAC) value of 0.2 mg/L for drinking waters (Table 1).

**Table 1.** *National and international water quality standards for aluminium (all concentrations in mg/L).*

| Organisation/<br>government | Guideline | MAC [*] | Reference |
|---|---|---|---|
| WHO | 0.2 | - | WHO (1984) |
| EEC | 0.05 | 0.2 | Council European Communities (1980) |
| Belgium | (0.05)[a] | 0.1 | Belgium Ministry of Health (1986) |
| FRG | (0.05)[a] | 0.2[b] | VEDEWA (1986) |
| Sweden | - | 0.15 | Sydvatten (1986) |
| Switzerland | 0.05 | 0.50 | Swiss Min.of Public Health (1986) |
| USA | 0.05[c] | - | American Public Health Assoc. (1976) |
| Finland[d] | - | 0.2 | Finnish Nat.Board of Health (1986) |
| Denmark[e] | 0.05 | 0.2 | Danish Min.of the Environment (1986) |
| Austria | - | 0.2 | Austrian Min.for Health and Environment (1986) |

[*] Maximum Acceptable Concentration
[a] EEC guideline
[b] Will refer to sum of natural content *and* treatment residuals as from 1/10/86 - previously excluded natural content
[c] American Public Health Association/American Water Works Association guideline (not legally enforceable)
[d] 35% of all drinking water treated using aluminium sulphate
[e] 2.5% max of supplies used aluminium sulphate coagulant

**Aluminium in West European drinking water**

*General*

A major aim of the survey was to present a clear picture of prevailing aluminium concentrations in potable water supplies in Western Europe and North America. Much of this information is widely scattered and has not previously been brought together in a report of this kind (Packham, 1986).

The following basic considerations were taken into account when conducting the survey:

(i)    Potable water is produced from a very wide variety of sources ranging from small borehole supplies to large upland impoundments.

(ii)   A range of treatment processes may be employed, depending on source, ranging from, for example, chlorination only for a borehole supply to more extensive treatment including coagulation, filtration, *etc.*, for a lowland river sources.

(iii)  Changes in water quality may occur within the distribution system itself. For example, traces of aluminium hydroxide floc, if not completely removed at the

**Table 2.** *Countries included in Western European drinking water survey.*

| Country | Population ($\times 10^6$) | % Population served by public water supply | Remarks |
|---|---|---|---|
| **EEC** | | | |
| Belgium | 9.85 | 97 | |
| Denmark | 5.1 | 85 | Very little alum used - mainly groundwater sources |
| FR Germany | 62.4 | 96 | |
| France | 54.3 | 97 | |
| Italy | 57 | 88 | |
| Netherlands | 14.1 | >99 | Very little alum used for for water treatment |
| United Kingdom | 55.9 | >99 | |
| **Non-EEC** | | | |
| Austria | 7.7 | 75 | Very little alum used for water treatment |
| Finland | 4.8 | 75 | |
| Norway | 4.1 | 75 | Coagulation rarely employed in water treatment |
| Sweden | 8.3 | 86 | |
| Switzerland | 6.4 | >99 | Very little alum used for water treatment |

treatment stage, may precipitate out in the distribution system itself (Knight, 1960), thus altering the final aluminium concentrations at the tap from that measured immediately after treatment.

(iv) It is only recently that aluminium has become a water quality parameter of interest, and has not therefore been a subject of routine monitoring in many water supply undertakings, especially in supplies where alum is not used as a coagulant. Because of this, data is limited in certain areas.

*Data acquisition*

Because of the extremely large geographical area covered by the boundaries of Western Europe, it was necessary to set reasonable limits on the number of countries included. Accordingly, twelve countries (Austria, Denmark, Belgium, Federal Republic of Germany (FRG), Finland, France, Italy, Netherlands, Norway, Sweden, Switzerland and the United Kingdom) were chosen, bearing in mind the following criteria:

**Table 3.** *Level of response to water undertaking enquiries.*

| Country | Water undertaking enquiries | |
|---|---|---|
| | Sent | Replies received |
| **EEC** | | |
| Belgium | 3 | 1 |
| Denmark | 1 | - |
| FR Germany | 52 | 16 |
| France | 24 | 16 |
| Italy | 12 | 6 |
| Netherlands | 1 | - |
| **Non-EEC** | | |
| Austria | 3 | - |
| Finland | - | - |
| Norway | 1 | 1 |
| Sweden | 4 | 4 |
| Switzerland | 26 | 13 |
| Total | 127 | 59 |

(i) Availability of details of the relevant water industries.

(ii) Likely availability of necessry data from water supply authorities.

Basic relevant statistics concerning these countries are given in Table 2.

The primary approach to data acquisition was to send letters of enquiry to as many water undertakings as possible. In addition to enquiries to individual water undertakings, enquiries concerning the wider national situation concerning aluminium in potable supplies were also sent to the relevant government authority in each of the countries concerned.

Table 3 summarises the level of response from water undertakings.

*Data presentation*

In the presentation of data which follows, it is considered that sufficiently wide variety of source types and treatment practice has been covered to give a clear picture of generally prevailing aluminium concentrations in treated water in many major Western European supplies. It should be noted that in all cases (except where otherwise stated), all potable water data refers to water sampled immediately after final treatment at the works concerned.

Available data have been presented in the following sections according to country of origin. In general, all aluminium concentrations given refer to total aluminium

**Figure 1.** *Water Authority boundaries in England and Wales.*

(dissolved and particulate) as determined by a spectrophotometric or atomic absorption method. The concentration data presented are usually expressed in terms of annual mean, maximum and minimum values except where otherwise indicated. Annual means are usually derived from at least one sample per month.

**Table 4.** *General data for English and Welsh Water Authorities, 1985.*

| Region | Area km$^2$ | Resident population (mid-1984) x10$^3$ | Average daily quantity supplied Water Auth. ML/day | Water Company ML/day | Total ML/day | Proportion of total supply Ground sources % | Surface sources % |
|---|---|---|---|---|---|---|---|
| Anglian | 27,358 | 5,114 | 1,100 | 557 | 1,657 | 45 | 55 |
| Northumbrian | 9,274 | 2,632 | 621 | 398 | 1,019 | 8 | 92 |
| North West | 14,445 | 6,870 | 2,525 | - | 2,525 | 11 | 89 |
| Severn Trent | 21,600 | 8,258 | 1,936 | 411 | 2,347 | 56 | 44 |
| Southern | 10,552 | 3,912 | 694 | 569 | 1,263 | 72 | 28 |
| South West | 10,884 | 1,435 | 434 | - | 434 | 12 | 88 |
| Thames | 13,100 | 11,575 | 2,618 | 1,194 | 3,812 | 4 0 | 60 |
| Welsh | 21,262 | 3,039 | 1,077 | 75 | 1,152 | 6 | 94 |
| Wessex | 9,918 | 2,366 | 386 | 499 | 885 | 50 | 50 |
| Yorkshire | 13,503 | 4,557 | 1,361 | 49 | 1,410 | 25 | 75 |
| England and Wales | 152,000 | 49,760 | 12,752 | 3,752 | 16,504 | 33 | 67 |

ML = megalitres

## United Kingdom

*General*

The water supply industries in Scotland and Northern Ireland are organised completely separately from those in England and Wales, and have not been included in the present survey. England and Wales have ten regional water authorities (WA's) whose boundaries correspond to the major river basins; these are shown in Figure 1. General data for all ten authorities is given in Table 4, detailed data is presented for five of these.

*(a) Northumbrian Water Authority*

Potable supplies within this authority's area (see Figure 1) are drawn from a combination of reservoir, river, spring and borehole supplies (Table 4). The data (Table 5) indicates a trend to slightly higher residual aluminium levels in the alum-using works. In order to clarify this, means of the average aluminium concentrations for the output of the alum-using works and the other works have both been calculated and compared on the basis of a simple statistical t-test (Table 5). The alum-using mean (0.12) is about twice as high as the mean from the non-alum users

208

**Table 5.** *Raw and potable water aluminium concentration data for surface sources in the Northumbrian WA area.*

| Daily Works output No. | ML/day | Source type | Coagulant used | Al concentration (mg/L) | | | | | | Mean pH (potable) |
|---|---|---|---|---|---|---|---|---|---|---|
| | | | | Max | Potable Min | Mean | Max | Raw Min | Mean | |
| 1 | 15.8 | Reservoir | Alum | 0.37 | <0.01 | 0.12 | na | na | na | 7.3 |
| 2 | 13.0 | Reservoir | Alum | 0.25 | 0.07 | 0.14 | 0.04 | 0.04 | 0.15 | 9.0 |
| 3 | 40.0 | River | Alum | 0.16 | 0.04 | 0.09 | na | na | na | 7.6 |
| 4 | 108.0 | River | Alum | 0.16 | 0.06 | 0.11 | na | na | na | 7.9 |
| 5 | 4.5 | Reservoir | Alum | 0.22 | 0.08 | 0.12 | 0.19 | 0.09 | 0.13 | 8.0 |
| 6 | 3.0 | Reservoir | Alum | 0.39 | 0.05 | 0.13 | 0.88 | 0.10 | 0.27 | 7.6 |
| 7 | 15.8 | Reservoir | Ferric sulphate | 0.34 | <0.01 | 0.10 | 0.27 | 0.09 | 0.18 | 8.0 |
| 8 | 53.1 | Reservoir | Chlor.copperas | 0.16 | <0.01 | 0.05 | na | na | na | 9.0 |
| 9 | 40.6 | Reservoir | Chlor.copperas | 0.12 | 0.03 | 0.06 | na | na | na | 9.2 |
| 10 | 108.0 | Reservoir | (a) Chlor.copperas (b) none (SSF) | 0.12 | 0.05 | 0.06 | na | na | na | 8.8 |

SSF = Slow sand filters
Average mean Al concentration (alum treated potable water) = 0.12 mg/L
Average mean Al concentration (no alum) = 0.07 mg/L
Difference statistically significant at $p > 0.99$
ML = megalitres

209

(0.07), and the difference is statistically significant at p > 0.99, indicating a possible trend to higher aluminium levels in alum-treated final waters. Both mean values are however well below the EEC MAC value of 0.2 mg/L, although some maximum values for individual works are above this. For those alum-treated supplies where raw water aluminium data are available, it is noticeable that mean treated aluminium values for these works are, in general, marginally lower than those found in the raw water, suggesting that a fraction of the natural aluminium content of the raw water is removed during treatment.

### (b) North West Water Authority

The majority (89%) of the authority's supplies are obtained from rivers and upland impoundments, the remainder originating from underground sources.

Inspection of the data (Table 6) shows that there is a slight trend to higher mean aluminium concentrations in alum-treated finished waters. A statistical comparison as used for Northumbrian WA data of average mean values for finished water aluminium concentrations was carried out. The mean values (0.09 mg/L for alum-using plants, 0.07 mg/L for remainder) are only slightly different however, and the difference is not statistically significant at p > 0.9. The maximum values for alum-using plants are, however, generally significantly higher than for the non-alum plants. All mean values, with the exception of works no.5, are within the EEC MAC limit. Comparison between raw mean aluminium concentrations and corresponding potable data show a consistent trend to lower potable values, indicating, as with Northumbrian WA data, that some removal of natural aluminium content takes place during treatment. In addition, comparison of recent data with the relevant 1978 means indicates that there has been a slight downward trend in the intervening period, presumably at least partly due to improved monitoring and control techniques.

### (c) Yorkshire Water Authority

Surface water sources account for 75% of the total in the authority's area. Table 7 presents the data in terms of annual means of aluminium in treated waters for the periods shown.

Comparison of alum-using works with the remainder makes it clear that aluminium levels in finished waters in the former are substantially higher than the latter. In five out of six cases the annual means are above the EEC 0.2 mg/L limit. Some of these high values are attributed by the authority to fluctuations in raw water quality but the performance of other plants where greater control of alum dosing should be possible is a cause for concern (Yorshire Water Authority, 1986) and it is evident that full compliance with the EEC MAC standard would appear to involve a considerable up-grading of monitoring and control procedures, investment in which is already proceeding.

### (d) Thames Water Authority

The Authority (including water companies) serves a population of 11.6 million, including all the central London area and the majority of its suburbs. Surface sources provided 60% of supply. Available data have been presented in Table 8.

**Table 6.** *Summary of Al concentration data for large surface water supplies in the North West WA area.*

| Works No. | Daily output ML/day | Source type | Coagulant used | Al concentration (mg/L) | | | | | | | Mean pH (Potable) |
|---|---|---|---|---|---|---|---|---|---|---|---|
| | | | | Max | Potable Min | Mean | Mean 85-86 | Max | Raw Min | Mean | |
| 1 | 245.2 | River | Alum | 0.38 | <0.01 | 0.03 | 0.05 | 1.0 | 0.04 | 0.10 | 8.3 |
| 2 | 80.0 | Reservoir | Alum | 0.52 | 0.02 | 0.12 | na | 0.28 | 0.02 | 0.13 | 8.6 |
| 3 | 72.6 | River | Alum | 0.28 | 0.01 | 0.05 | 0.05 | na | na | na | 8.0 |
| 4 | 45.0 | Reservoir | Alum | 0.32 | <0.01 | 0.03 | 0.02 | 0.27 | 0.01 | 0.06 | 8.0 |
| 5 | 40.5 | Reservoir | Alum | 0.56 | 0.06 | 0.24 | na | na | na | na | 8.4 |
| 6 | 102.3 | Reservoir | None | 0.16 | 0.07 | na | na | na | na | na | 8.0 |
| 7 | 36.0 | Reservoir | None | 0.22 | <0.02 | 0.09 | 0.06 | na | na | na | 8.2 |
| 8 | 15.0 | Lake | None | 0.07 | <0.02 | 0.03 | na | na | na | na | 8.6 |
| 9 | 12.5 | Reservoir | None | 0.24 | 0.02 | 0.10 | na | 0.49 | 0.04 | 0.22 | 7.7 |

Note: All data refer to 1978 except where stated
Average mean Al concentration (alum treated potable water) = 0.09
Average mean Al concentration (no alum) = 0.07
Difference not significant at $p > 0.9$
ML = megalitres

**Table 7.** *Mean annual aluminium concentrations (1982-85) in potable water from ten largest surface water plants in Yorkshire WA area.*

| Works No. | Coagulant | Mean annual Al concentration (mg/L) | | | |
|---|---|---|---|---|---|
| | | Potable | | Raw | used |
| | | 82-83 | 83-84 | 84-85 | 84-85 |
| 1 | Alum | 0.12 | 0.12 | 0.14 | na |
| 2 | Alum | 0.16 | 0.16 | 0.20 | na |
| 3 | Alum | 0.29 | 0.20 | 0.23 | na |
| 4 | Alum | 0.24 | 0.28 | 0.25 | 0.48 |
| 5 | Alum | 0.35 | 0.38 | 0.44 | 0.38, 0.5[*] |
| 6 | Alum | 0.43 | 0.62 | 0.73 | 0.32 |
| 7 | Fe(III)sulphate | na | na | 0.06 | na |
| 8 | None[a] | 0.12 | 0.12 | 0.15 | 0.22 |
| 9 | None[a] | na | na | 0.06 | na |
| 10 | None[a] | na | na | 0.02 | na |

[*] two sources.
[a] slow sand filters.
pH range of water in distribution (entire YWA area): 6.5-9.5
   (98% of samples).
Average mean potable Al concentration (alum-treated) = 0.33
Average mean potable Al concentration (no alum)    = 0.07
Difference is statistically significant at $p > 0.95$

The table indicates clearly that the majority of water going into supply from these sources is well within the EEC MAC limit for aluminium. Comparison of the 1983 aluminium means (potable) with the 1978 equivalents indicates that, in general, no significant change has occurred in average concentrations in the intervening period. Comparison with raw means shows that no distinct trend in the effect of alum-based coagulants in increasing finished water aluminium concentrations compared with raw is evident.

*(e) South West Water Authority*
The South West WA takes 88% of its supply from surface waters. Table 9 presents aluminium data for eight of the largest supplies in the area.

For the alum-treated supplies (works 1-5), there is generally no significant increase in aluminium concentration between raw and potable waters. Raw water aluminium levels for works no.3 have risen noticeably in the period 1984-86, a trend in tandem with a lowering of raw water pH values, but potable aluminium concentrations appear not to have been greatly affected. In general, all supplies in the

**Table 8.** *Aluminium concentration data for Thames WA area, 1982-83.*

| Works No. | Coagulant | Al concentration (mg/L) | | | | | | | | Mean pH (potable) |
|---|---|---|---|---|---|---|---|---|---|---|
| | | Potable | | | | Raw | | | | |
| | | Max | Min | Mean | (1978 Mean) | Max | Min | Mean | (1979 Mean) | |
| 1 | Alum | 0.13 | 0.04 | 0.08 | (0.06) | 0.02 | <0.01 | <0.01 | (0.02) | 7.7 |
| 2 | Alum | 0.24 | <0.01 | 0.08 | (0.10) | 0.02 | <0.01 | <0.01 | (0.02) | 7.7 |
| 3 | Alum | 0.18 | <0.01 | 0.08 | (0.08) | na | na | na | ( na ) | 7.7 |
| 4 | Alum | 0.40 | 0.04 | 0.04 | (0.04) | 0.02 | <0.01 | <0.01 | (0.04)* | 7.6 |
| 5 | Alum | 0.10 | 0.01 | 0.05 | (0.04) | 0.03 | <0.01 | 0.01 | (0.02 ) | 7.6 |
| 6 | Alum | 0.05 | <0.01 | 0.02 | (0.04) | 0.09 | <0.01 | 0.04 | ( na ) | 7.4 |
| 7 | Alum | 0.13 | <0.01 | 0.06 | (0.08) | 0.07 | 0.02 | 0.05 | (0.01)* | 7.6 |
| 8 | Alum | 0.06 | <0.01 | 0.02 | (0.25) | na | na | na | (0.12) | 8.1 |
| 9 | Alum | 0.12 | 0.01 | 0.06 | (0.07) | na | na | na | ( na ) | 7.4 |
| 10 | Alum | 0.09 | <0.01 | 0.02 | (0.03) | 0.20 | <0.01 | <0.01 | (0.05) | 7.6 |
| 11 | Fe(III)sulphate | 0.03 | <0.01 | <0.01 | (<0.01)* | 0.28 | <0.01 | <0.01 | (0.03) | 7.8 |

All raw water for these works is taken from rivers
* single source

**Table 9.** *Aluminium data (1981-86) for large surface and groundwater supplies in South West WA.*

| Works No. | Daily output ML/day | Source type | Coagulant used | Mean aluminium concentrations (mg/L) | | | | | Range of raw means 1981-1986 | Range of raw annual pH means 1981-1986 |
|---|---|---|---|---|---|---|---|---|---|---|
| | | | | 81/82 | Potable 82/83 | 83/84 | 84/85 | 85/86 | | |
| 1 | 32.3 | River | Alum | 0.05 | 0.06 | 0.07 | 0.07 | na | 0.03-0.14 | 6.62-7.80 |
| 2 | 25.7 | River | Alum | 0.12 | 0.05 | 0.10 | 0.10 | 0.08 | 0.03-0.20 | 7.05-7.16 |
| 3 | 25 | River/Res. | Alum | 0.12 | 0.08 | 0.23 | 0.23 | 0.10 | 0.10-7.68* | 3.90-7.00[a] |
| 4 | 22.5 | Reservoirs | Alum | 0.16 | 0.15 | 0.23 | 0.23 | 0.15 | 0.13-0.20 | 6.17-6.90[b] |
| 5 | 14 | Reservoir | Alum | na | na | 0.09 | 0.09 | 0.12 | 0.10-0.12[c] | 5.5-8.0 |
| 6[d] | 5.6 | Groundwater | None | | 0.01 - 0.03 | | | | 0.01 or less | 7.20-7.90[a] |
| 7 | 5.2 | Groundwater | None | 0.14 | 0.16 | 0.17 | 0.14 | 0.18 | 0.10-0.20 | 5.76-7.07 |
| 8 | 0.17 | Groundwater | None | 0.19 | 0.21 | 0.21 | 0.23 | 0.23 | 0.19-0.26 | 4.90-5.41 |

[a] data from single samples
[b] data for 1981-84
[c] data for 1983-86
[d] data for total supply from 4 boreholes
* = mean of 3 samples
ML = megalitres

1985-86 period have met the 0.2 mg/L EEC standard, except works 8, and it is noteworthy that the alum- treated surface waters for the same period are generally lower than groundwater output from works 7 and 8.

**Federal Republic of Germany**

Replies were received from undertakings supplying greater than 15% of the total FRG population, in both northern and southern regions. Aluminium concentrations in potable supplies for about half of these supplies are generally too low to warrant routine analysis, and no detailed data are available. Data for the remaining undertakings are summarised in Table 10.

Data from the Göttingen/Hannover area indicate clearly that potable water aluminium concentrations fall within the range 0.018-0.062 mg/L limit. In addition, only the mean value from works 3 (0.062 mg/L is above the EEC 0.05 mg/L guideline value, while all maximum values are within the 0.2 mg/L limit. The mean potable concentrations are also in all cases lower than the corresponding raw water values.

Data from the suppliers to the Frankfurt area (Table 10) indicate that here also prevailing concentrations are very satisfactory both with respect to the 0.2 and 0.05 mg/L limits. The data presented were stated to be values from three typical plants out of a total of more than 20 under the authority's management. Unfortunately, the type of source or treatment process pertaining to the three works in Table 10 was not stated, but from comparison with other data in the table it appears likely that works 3 uses a groundwater source, in view of the very low aluminium concentrations. Data from the Stuttgart region indicates that the latter values are generally lower still, and emphasises the general trend to very low aluminium concentrations in most groundwater derived supplies.

The remaining data in Table 10 covers a total of 23 works. Maximum values indicate that EEC guideline and MAC values are rarely exceeded, but sufficient information was not available to examine trends more closely.

**Italy**

Data was received from four undertakings, and these are summarised in Table 11. In general, nearly all mean and maximum values are within the EEC 0.2 mg/L limit. A slight trend towards higher aluminium concentrations in potable water than in raw is also noticeable.

**Sweden**

Information was received from undertakings supplying the four major population centres. All the producers used alum as a coagulant, and their data cover supplies to nearly 30% of the total Swedish population. Detailed aluminium data are given in Table 12.

**Table 10.** *Aluminium data from West German waterworks.*

| Region/ Works No. | output ML/day | Source type | Daily Coagulant used | Al concentration (mg/L) Potable Max | Min | Mean | Raw Max | Min | Mean | Mean pH Potable | Raw |
|---|---|---|---|---|---|---|---|---|---|---|---|
| **Göttingen/Hannover** | | | | | | | | | | | |
| 1 | 126.0 | Upland impoundment | Alum | 0.036 | 0.011 | 0.018 | 0.220 | 0.026 | 0.083 | 8.54 | 7.14 |
| 2 | 38.4 | " | Alum | 0.051 | 0.023 | 0.033 | 0.930 | 0.680 | 0.844 | 9.13 | 4.76 |
| 3 | 47.1 | " | Alum | 0.040 | 0.014 | 0.021 ) | 0.135 | 0.046 | 0.083 | 9.03 ) | 6.66 |
| 4 | 32.9 | " | Alum | 0.044 | 0.029 | 0.062 ) | | | | 8.88 ) | |
| **Stuttgart/Ulm** | | | | | | | | | | | |
| 1 | na | Lake Constance | None | 0.004 | <0.001 | 0.001 | 0.003 | <0.001 | 0.002 | 7.89 | 8.35 |
| **Frankfurt** | | | | | | | | | | | |
| 1 | na | Not stated | Not stated | na | na | 0.040 | na | na | 0.038 | 7.84 | 7.06 |
| 1 | na | Not stated | Not stated | na | na | 0.016 | na | na | 0.017 | 8.02 | 7.00 |
| 2 | na | Not stated | Not stated | na | na | 0.004 | na | na | 0.005 | 7.22 | 7.21 |
| **Ruhr (all data covers 22 works)** | | | | | | | | | | | |
| - | | Surface/Ground | Not stated | Range of individual samples: <0.01-0.24 | | | | | | | |
| | | | | Range of annual geometric means (1983-85): <0.01-0.05 | | | | | | | |
| **Freiburg** | | | | | | | | | | | |
| - | | Not stated | None | Range of six individual samples (1982-85): 0.00-0.116 | | | | | | | |
| **Swiss Border** | | | | | | | | | | | |
| - | | Lake Constance | None | "Mean value": <0.001 | | | | | | | |

na = not available
ML = megalitres

216

**Table 11.** *Aluminium data from Italian waterworks.*

| Region/ Works No. | Daily output ML/day | Source type | Coagulant used | Al concentration (mg/L) | | | | | | Mean pH | |
|---|---|---|---|---|---|---|---|---|---|---|---|
| | | | | Potable | | | Raw | | | | |
| | | | | Max | Min | Mean | Max | Min | Mean | Potable | Raw |
| **Rome** | 64.8 | River | PAC[*] | 0.085 | 0.041 | 0.061[a] | 0.021 | 0.014 | 0.017[a] | 8.9 | 8.0 |
| **Genoa** | | | | | | | | | | | |
| 1 | 31 | River | Alum | 0.019 | <0.01 | 0.015 | na | na | <0.01 | 7.8 | 8.1 |
| 2 | 80 | Reservoir | Alum | 0.087 | <0.01 | 0.022 | 0.057 | <0.01 | 0.019 | 7.8 | 7.9 |
| 3 | 28 | Reservoir | Alum | 0.125 | <0.01 | 0.088 | 0.057 | <0.01 | 0.026 | 7.6 | 7.6 |
| **Turin** | | | | | | | | | | | |
| 1 | ) | River Po | PAC[*] | 0.200 | 0.050 | 0.150 | na | na | na | 7.6 | na |
| 2 | ) | River Po | Not stated | 0.010 | 0.002 | 0.005 | na | na | na | 7.6 | na |
| 3 | )80 | River Po | Not stated | 0.040 | 0.002 | 0.010 | na | na | na | 7.5 | na |
| 4 | )total | River Po | Not stated | 0.040 | 0.004 | 0.010 | na | na | na | 7.3 | na |
| 5 | ) | River Po | Not stated | 0.020 | 0.003 | 0.005 | na | na | na | 7.1 | na |
| 6 | ) | River Po | Not stated | 0.010 | 0.002 | 0.005 | na | na | na | 7.6 | na |
| **Florence** | | | | | | | | | | | |
| 1 | 346 | River | Alpoclar/Prodefloc | 0.33 | 0.01 | 0.09 | 19.00 | 0.24 | 3.09 | 7.5 | 8.0 |
| 2 | 26 | River | Prodefloc | 0.07 | 0.00 | 0.02 | 7.5 | 0.35 | 2.38 | 7.3 | 7.9 |

[a] covers an 18 month period Jan 1985 - July 1986

[*] PAC = polyaluminium chloride

ML = megalitres

The table shows that all mean annual aluminium concentrations and associated maxima are well within the Swedish recommended maximum of 0.15 mg/L (Table 1) and the EEC MAC value, and, in general, these values are amongst the lowest for all the Western European alum-treated potable water plants for which data are available. Where raw water data are available, there is a clear trend to treated concentrations being lower than those for raw.

## Switzerland

Data are available for seven undertakings, summarised in Table 13.

Considering data for the supplies abstracted from Lake Geneva first, it is clear that the annual means for potable water are well within the EEC 0.2 mg/L limit, although Switzerland is not an EEC member and is not obliged to meet these limits. A national guideline limit of 0.05 mg/L exists at present, however, with an upper limit of 0.5 mg/L. With respect to the guideline 0.05 mg/L value, therefore, the mean aluminium concentrations in potable water from the Geneva supplies appears slightly high, but is probably not considered worthy of very serious concern in view of the much higher upper limit in force.

Data for supplies abstracted from Lake Zurich fall into two main groups, the first three works supplying Zurich itself (under a single authority), and the remainder supplying smaller areas in the same region. Figures from works 1, 2 and 3 show predictably lower concentrations in the groundwater supplies from works 3 than for the lake-derived supplies. It is also interesting to note from the data as a whole that, in addition to the use of alum, a number of plants use polyaluminium chloride (PAC), a polymeric aluminium salt, and data from these two sources indicate that significantly lower aluminium concentrations in finished waters were obtained.

In addition to data from these two areas, more general data for raw and treated waters were made available by the Swiss Ministry of Public Health (personal communication, 1986), and is tabulated in Table 14. From this it can be seen that, once again, ground and spring water aluminium concentrations are generally much lower than treated (usually alum) lake water which makes up 27% of total drinking water supplies. Although the table only shows typical values, the mean value of 0.145 mg/L for alum-treated water perhaps indicates that the application of a 0.05 mg/L limit would pose problems for alum-using plants.

## Other European countries

Less information was forthcoming from the remaining six countries. The majority of replies from France indicated that groundwater sources were widely used, which required no coagulation. In other cases, only limited data were available which indicated that aluminium residuals were generally less than 0.1 mg/L. No data was received from Austria or Finland. Relevant data received from Belgium (Antwerp), Denmark (Copenhagen), the Netherlands (Rotterdam) and Norway (Oslo) are summarised in Table 15. Residuals for Copenhagen, Rotterdam and Oslo are clearly very low, and in the latter two cases this is probably largely due to use of ferric

Table 12. *Aluminium data from Swedish waterworks.*

| City/ Works No. | Daily output ML/day | Source type | Coagulant used | Al concentration (mg/L) | | | | | | Mean pH | |
| --- | --- | --- | --- | --- | --- | --- | --- | --- | --- | --- | --- |
| | | | | Potable | | | Raw | | | Potable | Raw |
| | | | | Max | Min | Mean | Max | Min | Mean | | |
| **Malmo** | | | | | | | | | | | |
| 1 | 104 | Lake | Alum | 0.05 | <0.05 | <0.05[b] | na | na | 0.13 | 8.1[b] | 7.9 |
| **Göteborg** | | | | | | | | | | | |
| 1 | 88 | Lowland River | Alum | na | na | <0.05 | na | na | 0.15 | 8.5 | 7.2 |
| 2 | 97 | Lakes[a] | Alum | na | na | <0.05 | na | na | 0.08 | 8.5 | 7.2 |
| **Stockholm** | | | | | | | | | | | |
| 1 | na | Surface | Alum | 0.127 | 0.016 | 0.055 | na | na | na | 8.5 | 7.6 |
| 2 | 177 | Lake | Alum | 0.08 | <0.02 | 0.04 | na | na | na | 8.5 | na |

[a] used as storage for water from lowland water source
[b] median values

na = not available
ML = megalitres

**Table 13.** Aluminium data from Swiss waterworks.

| Region/ Works output No. | ML/day | Daily Source type | Coagulant used | Al Concentration (mg/L) Potable Max | Min | Mean | Raw Max | Min | Mean | Mean pH Potable | Raw |
|---|---|---|---|---|---|---|---|---|---|---|---|
| **Lake Geneva** | | | | | | | | | | | |
| 1 | 129.6 (max) | Lake Geneva | Alum | 0.215 | 0.042 | 0.078 | 0.036 | 0.000 | 0.011 | 7.99 | 7.94 |
| 2 | 38.9 (max) | Lake Geneva | Alum | 0.105 | 0.034 | 0.063 | 0.018 | 0.000 | 0.009 | 7.95 | 7.89 |
| 3* | | Lake Geneva | PAC** | 0.09 | 0.04 | 0.05 | 0.04 | 0.01 | 0.02 | 7.95 | 7.97 |
| 4 | | Lake Geneva P | AC** | 0.05 | 0.03 | 0.04 | 0.06 | 0.02 | 0.03 | 7.74 | 7.96 |
| **Lake Zurich** | | | | | | | | | | | |
| 1 | ) 133.3 | Lake Zurich (72%) | Alum | 0.055[b] | 0.025[b] | 0.035[b] | 0.050[a] | 0.010[a] | 0.035[a]) | 7.93 | 7.86 |
| | ) | Spring water (28%) | None | na | na | na | 0.005 | <0.005 | 0.015 ) | | 7.58 |
| 2 | ) | Lake Zurich | Alum | 0.300 | 0.105 | 0.170 | 0.070 | 0.005 | 0.040 | 8.06 | 7.85 |
| 3 | 40.2 | Groundwater | None | 0.040 | 0.005 | 0.010 | 0.035 | 0.005 | 0.015 | 7.91 | 7.71 |
| 4 | na | Lake Zurich | Alum | na | na | 0.16 | na | na | na | na | na |
| 5 | na | Lake Zurich | PAC** | na | na | 0.03 | na | na | na | na | na |
| 6 | na | Lake Zurich | PAC** | na | na | 0.04 | na | na | na | na | na |
| 7 | na | Lake Zurich | PAC** | na | na | 0.04 | na | na | na | na | na |
| 8 | 3.0 | Lake Biel | Alum | 0.16 | 0.02 | 0.05 | 0.14 | 0.00 | 0.05 | 8.29 | 7.97 |

* Data for 1985 and 1986
** PAC = polyaluminium chloride
[a] chlorinated raw water
[b] after mixing with spring water
na = not available
ML = megalitres

**Table 14.** *Typical aluminium concentrations in Swiss Waters (Swiss Ministry of Public Health, 1986).*

| Type of Water | Al Concentration (mg/L) Range | Mean[a] |
|---|---|---|
| Raw lake water | 0.005-0.07 | 0.030 |
| Treated lake water | 0.055-0.300 | 0.145 |
| Groundwater | 0.005-0.035 | 0.010 |
| Spring water | 0.005-0.065 | 0.015 |

[a] relevant number of data points not stated.
Lake water in Switzerland supplies 27% of drinking water.
Aluminium-based coagulants are only used for treating lake water.
Swiss guideline value for potable water aluminium concentration is 0.05 mg/L, MAC is 0.5 mg/L.

sulphate (Rotterdam) or no coagulant (Oslo). The annual means for the Antwerp data show a downward trend for the years 1983-85, which reflects the imposition of a legal limit of 0.1 mg/L for aluminium in drinking water since April 1984 (Belgian Ministry of Health, personal communication, 1986).

**United States**

Although the scope of this survey centred upon West European data, useful comparisons can be made with the work of Miller *et al.* (1984) who described the results of a nationwide EPA- sponsored survey of the occurrence of aluminium in raw and treated waters.

The survey was based on ten regions of the US, based on 290 different water supply systems from which a total of 1,517 individual samples were analysed for iron and aluminium. A major aim of the survey was to detect differences in aluminium residuals between alum-using and other types of treatment plants. The study concluded that when alum was used as a coagulant, there was a 40-50% chance that the concentration of aluminium will increase above the original concentration in raw water. Where iron was used as a coagulant, concentrations of aluminium (raw and finished) decreased, but the number of plants using iron was relatively small (19 *c.f.* 75 using alum). The concentrations of aluminium in finished waters were not related with either the size of the plant studied or with the region of location.

In terms of broad comparisons with European data, this study indicates clearly that maximum aluminium concentrations are regularly in excess of those found in the European survey. This may partly reflect the absence of a Federal limit on drinking

**Table 15.** *Data for Belgium, Denmark, Netherlands and Norway.*

| Location/ Works No. | Daily output ML/day | Coagulant used | Al concentration in potable supplies (mg/L) | | | | | | | | | Mean pH (1985) | |
|---|---|---|---|---|---|---|---|---|---|---|---|---|---|
| | | | 1983 | | 1984 | | | 1985 | | | Mean | Potable | Raw |
| | | | Max | Min | Mean | Max | Min | Mean | Max | Min | | | |
| **Belgium (Antwerp)*** | | | | | | | | | | | | | |
| 1 | 135 | Alum | 0.181 | 0.020 | 0.088 | 0.140 | <0.001 | 0.050 | 0.147 | <0.001 | 0.038 | 7.94 | 8.24 |
| 2 | 400 | Alum** | 0.262 | 0.005 | 0.146 | 0.260 | 0.008 | 0.070 | 0.156 | 0.008 | 0.057 | 5.99 | 8.27 |
| **Denmark (Copenhagen)**** | | | | | | | | | | | | | |
| - | na | Alum | na | na | na | na | na | na | 0.05 | 0.01 | 0.03 | na | na |
| **Netherlands (Rotterdam)** | | | | | | | | | | | | | |
| 1 | na | Ferric sulphate | na | na | na | na | na | na | 0.007 | <0.001 | 0.003 | 8.2 | 9.0 |
| 2 | na | Ferric sulphate | na | na | na | na | na | na | 0.011 | 0.002 | 0.004 | 8.25 | |
| **Norway (Oslo)** | | | | | | | | | | | | | |
| a | - | None | na | na | na | na | na | na | 0.16 | 0.07 | 0.08 | na | na |
| a | - | None | na | na | na | na | na | na | 0.09 | 0.03 | 0.07 | 6.38 | 6.40 |
| a | - | None | na | na | na | na | na | na | 0.09 | 0.02 | 0.07 | na | na |

\* population supplied: 1 million

\*\* occasional surface supply - 97.5% of annual supplies from groundwater using no coagulant

a data from three points in distribution system only

ML = megalitres

water aluminium content, in contrast to the EEC limit, which appears to serve as a working guideline even for water suppliers in European countries which are not EEC members.

## Overview of Recent Research on Aluminium in Potable Waters

The overall impression gained from a comprehensive overview of recent and current research on the aluminium/drinking water topic in Western Europe carried out as part of this survey is that, not unexpectedly, most concern on the topic is evident in countries such as Sweden, Norway and the USA where some potable water sources, especially groundwater, have been seriously affected by acidification and consequent increased in trace metal contamination, including aluminium. The contribution to the aluminium content of treated waters arising from the use of aluminium-based coagulants has not, in contrast, received as much attention to date, except in France, where considerable research interest is evident. Interest in wider environmental effects of acidification and aluminium levels in natural waters is more widely spread, but is outside the scope of this paper.

Research is concentrating on three broad areas: optimisation of existing treatment processes, use of alternative aluminium based flocculants, and new processes. Most work is being conducted on the first two options. In Germany and France in particular the optimisation of flocculation is being investigated (Clement et al., 1986; Guilini Chemie, 1986). In addition, flocculants such as polyaluminium chloride (PAC) are attracting considerable interest. In comparison to aluminium sulphate, PAC gives faster and more effective flocculation event in cold water conditions with subsequent lower aluminium residuals.

In Scandinavia there is increasing concern about levels of aluminium in groundwater supplies, and the difficulties of overall assessment of acidification effects on public water supplies in Sweden have been investigated (Grimvau et al., 1986) but no specific attention was given to aluminium. More pertinently, Anderson and Hedberg (1985) of the Chalmers University of Technology have published details of pilot studies on aluminium removal plants for use in individual dwellings relying on groundwater supplies. Filtration over a variety of media was examined (dolomite, calcium carbonate, etc.) as well as ion exchange, reverse osmosis and other techniques, and satisfactory aluminium removal was found with a number of these techniques. Close control of pH in public supplies has also been examined at this Institute (Hernebring, 1980).

## Comparative Human Aluminium Intake from Water and Food

Any assessment of human exposure to aluminium through consumption of drinking water must be put in the context of the additional daily ingestion of aluminium in food and beverages. An extensive study by the UK Ministry of Agriculture, Fisheries and Foods [MAFF] (1985) analysed the aluminium content of diets and typical adult humans in the UK, and derived estimated daily intake values (Tables 16 and 17). On the basis of this study, the drinking water element of the daily intake was found to

**Table 16.** *Aluminium content of some typical fresh vegetables and other foods (from MAFF, 1985).*

| Food or beverage | No. of samples | Range | Mean |
|---|---|---|---|
| Broad beans | 5 | 4.3-11.9 | 6.4 |
| Brussels sprouts | 5 | 2.2-70 | 14.5 |
| Cabbage | 5 | 0.6-2.9 | 1.6 |
| Carrots | 5 | 15.2-40.5 | 26.0 |
| Celery | 5 | 1.3-11.3 | 5.9 |
| Lettuce | 5 | 6.2-810 | 234 |
| Parsnips | 5 | 6.0-82 | 39.1 |
| Potatoes | 5 | 2.9-13.7 | 8.0 |
| Tomatoes | 5 | 10.1-32.8 | 21.5 |
| Lager[*] | 4 | 0.2-0.3 | 0.2 |
| Bitter Lemon[*] | 1 | - | 1.3 |
| Paté[*] | 7 | 0.6-1.8 | 1.3 |
| Ham | 3 | 0.2-0.9 | 0.5 |
| Sardines in tomato sauce[*] | 2 | - | 0.1 |
| Salmon | 1 | - | 1.9 |

[*] Food in aluminium cans

range from 0.5% to 11%. From the potable water data presented in previous sections, however, the great majority of annual mean aluminium concentrations found were well below the quoted MAFF value of 0.48 mg/L, and in general it appears that, on the basis of the MAFF study, drinking water in Western Europe would usually contribute 5% or less of the total daily aluminium intake.

Other studies of total dietary aluminium intake noted in the MAFF report have, however, indicated a fairly wide range of estimated daily intake values, ranging from 7 mg/day (Varo and Koivistoinen, 1980) to 36.4 mg/day (Kehor et al., 1940). In addition, estimated daily intake data from American sources given by Epstein (1984) indicates that much higher values are possible, particularly if proprietary medicines containing high aluminium contents such as antacids are used regularly (Table 18). A noticeable feature of this data is the specific inclusion of data concerning aluminium content of food additives, which appears, apart from drug usage, to be a major source of aluminium intake. This aspect of diet was not specifically examined in the MAFF study. More pertinently, however, the estimated drinking water contribution quoted

**Table 17.** *Estimated daily aluminium intake of general UK population (from MAFF, 1985).*

| | mg Al/person/day | |
| | Minimum | Maximum |
| --- | --- | --- |
| **Contribution from food** | | |
| (Results of diet study) | 6 | 6 |
| **Contribution from air** | | |
| (Assuming Al concentrations vary from 58 to 208 mg/m$^3$, respiratory rate of 15 m$^3$/day, 24 hour exposure, all Al deposited in the lung) | $0.8 \times 10^{-3}$ | $3.1 \times 10^{-3}$ |
| **Contribution from water** | | |
| (Assuming mean consumption of 1.5 L/day, tap water concentrations between 0.02 and 0.48 mg/L) | $3.0 \times 10^{-2}$ | $7.2 \times 10^{-1}$ |
| Total intake | 6.03 | 6.72 |

by Epstein (1 mg/day) is very similar to the MAFF maximum figure of 0.72 mg/day. Bearing in mind that this is probably much higher than that normally found, it is apparent that drinking water makes a relatively insignificant contribution to daily aluminium intake when compared with other sources. A similar conclusion was also reached by McDonald (1985).

The relationship between daily intake and the amount of aluminium actually absorbed by the body is at present far from clear, and a comparison of daily intakes from drinking water and other sources must be undertaken in that light, especially since the precise nature of aluminium species in potable supplies is still the subject of research (World Health Organisation, 1984). In addition, behaviour of aluminium from a variety of sources under the conditions normally found in the human stomach and the subsequent effects on absorption are still very much subjects of debate.

### General Summary

(i)   Aluminium concentrations in natural surface and groundwaters of neutral pH are generally very low ( 0.05 mg/L), but levels may be elevated in regions where pH values have been lowered due to the effects of acidic deposition.

(ii)  From the survey of Western European drinking water aluminium concentrations, groundwater supplies in all countries replying generally showed very low values, although elevated concentrations were evident in some local small-scale supplies where acidification has taken place.

**Table 18.** *Estimated daily intake of aluminium from various sources (US data from Epstein, 1984).*

| Category | Source | Al (mg/day) |
|----------|--------|-------------|
| Food | Natural content | 3-10 |
|  | Intentional additives (FDA-approved Al compounds) | 25-50 |
|  | Unintentional additives (from metallic Al products) | 2.5 |
| Water | Natural content, alum | 1 |
| Air | Dust, smoke, toiletry, sprays | ? |
| Drugs | Antacids | 50-1,000+ |
|  | Buffered aspirin | 10-100+ |

FDA = Federal Drugs Administration.

(iii) Data for water treated with aluminium-based coagulants indicated that the majority of plants surveyed performed satisfactorily with respect to meeting the EEC 0.2 mg/L, in terms of annual mean concentrations. Exceptions were found in some areas of the UK where poorly-buffered river sources in some areas can be subject to large variations in raw water quality, sometimes due to the effects of acid deposition.

(iv) Although the 0.2 mg/L EEC standard was met satisfactorily in most cases, it appears that a reduction of the EEC MAC value to 0.05 mg/L could pose serious problems of compliance for producers in Belgium, France, Italy and the UK, and, outside the EEC, to some suppliers in Switzerland, if EEC trends were followed.

(v) A large-scale study of aluminium in drinking water in the US found that there was a 40-50% chance of potable supplies containing a higher aluminium concentration than the raw water when alum was used as a coagulant. The overall range of levels found was similar to that found in European data.

(vi) A survey of recent and on-going research in Western Europe and the USA indicates that aluminium in drinking water is a research topic of considerable interest in countries where elevated aluminium levels due to acidification in groundwaters have taken place, and is seen as a potential health problem. For larger water undertakings, considerable research interest is shown in improving the use of alum as a coagulant and developing new products.

(vii) Drinking water forms only a minor proportion ( 5%) of most adult daily aluminium intake. The relative contribution of food, water and other sources to ultimate aluminium absorption is a subject of considerable uncertainty at the present time.

# References

American Public Health Association (1976). *Standard Methods for the Examination of Water and Wastewater*, 16th Edition. Washington, DC.

Andersson, O. and Hedberg, T. (1985). Internal Report 315, Chalmers Technological University, Göteborg.

Austrian Ministry for Health and Environment, Vienna (1986). Personal Communication.

Belgian Ministry of Health, Brussels (1986). Personal Communication.

Cawse, P.A. (1974). *Atomic Energy Research Establishment Report* R7669. HMSO, London. (Quoted by Odonnell *et al.*, 10).

Clement, M. *et al.* (1986). *Tribune du Cebedeau*, **36**, no.480, 469.

Council of the European Communities (1980). Directive of 15 July 1980 relating to the quality of water intended for human consumption. 80/778/EEC *Off. J. European Communities* L229.

Danish Ministry of the Environment, Copenhagen (1986). Personal Communication.

Dickson, W. (1986). *Water Quality Bull.*, **11**(1), 39.

Dinman, B.D. (1984). *Aluminium and Health: Alzheimer's Disease*. International Primary Aluminium Institute, London.

Driscoll, C.T. and Newton, R.M. (1985). *Environ. Sci. Technol.*, **19**(11), 1019.

Durum, W.H. and Haffty, J. (1963). Implications of the minor element content of some major streams of the world. *Geochim. Cosmochim. Acta*, **27**, 1.

Edwardson, J.A. *et al.* (1986). *The Lancet*, 354.

Epstein, S.G. (1984). *Aluminium and Health*. The Aluminium Association, Washington, DC.

Finnish National Board of Health, Helsinki (1986). Personal Communication.

Grimvall, A. *et al.* (1986). *Water Quality Bull.*, **11**(1), 6.

Guilini Chemie GmbH, Hannover, FRG (1986). Internal Report.

Henriksen, H. and Kirkhusimo, L.A. (1986). *Water Quality Bull.*, **11**(1), 34.

Hernebring, C. (1980). Publication B80:1, Chalmers Technological University, Göteborg.

Hultberg, H. and Johansson, K. (1981). *Nordic Hydrology*, **21**, 51.

Kehoe, R.A. *et al.* (1940). *J. Nutrition*, **19**, 579.

Krishnan, S.S. and Crapper McLachlan (1985). *Science Total Environment*, **41**, 203.

Lewis, W.M. and Waddington, J.I. (1986). *Proc. Intl. Conf. on Chemicals in the Environment*, 7. Lisbon.

McDonald, M.E. (1985). *Environ. Sci. Technol.*, **19**(4), 772.

Miller, J.R. and Andelman, J.B. (1986). *Proc. Intl. Conf. on Chemicals in the Environment*, 26. Lisbon.

Miller, R.G. *et al.* (1984). *J. Am. Waterworks Assoc.*, **76**(1), 84.

Ministry of Agriculture, Fisheries and Food (MAFF) (1985). *Food Surveillance*. Paper No.15. HMSO, London.

Odonnell, A.R. *et al.* (1984). *Water Research Centre Technical Report* TR197.

Packham, R. (1986). Water Research Centre. Personal Communication.

Sung, W. *et al.* (1984). *J. New England Water Works Assoc.*, **98**(4), 363.

Swiss Ministry of Public Health, Bern (1986). Personal Communication.

Sydvatten, A.B., Malmo, Sweden. (1986). Personal Communication.

Talbot, R.W. and Elzerman, A.W. (1985). *Environ. Sci. Technol.*, **19**(6), 553.

Taylor, F.B. *et al.* (1986). *Water Quality Bull.*, **11**(1), 50.

Varo, P and Koivistoinen, P. (1980). *Acta Agric. Scand. Suppl.*, **22**, 165. (Quoted in 27).

VEDEWA (Water and Sewage Works Association), Stuttgart (1986). Personal Communication.

World Health Organisation (WHO) (1984). *Guidelines for Drinking Water Quality*, Vol.1. Geneva.

Yorkshire Water Authority (1986). *Water Quality 1984-1985.*

# 22 The Effects of Aluminium from Acid Waters on Aquatic Ecology and Drinking Water Supplies in Wales

A.P. Donald and L. Keil
*Welsh Water Authority, Penyfai House, Furnace, Llanelli, Dyfed SA15 4EL, Wales*

Although aluminium is a significant component of most rock types, it is only under certain environmental conditions that concentrations of the element become elevated in natural waters. Such conditions have been detected in the uplands of Wales, and result from acid deposition from industrial emissions falling on an area where the rocks and soils afford minimal buffering. The effects are enhanced where conifer afforestation has occurred (Stoner *et al.*, 1984).

In an investigation carried out throughout Wales mean rainfall pH was found to be as low as 4.3 in the uplands of Mid and North Wales (Donald and Stoner, in press). The lower Paleozoic shales and greywackes of this region are among the most base-poor lithologies in the British Isles. Streams and lakes throughout the "vulnerable" upland area were sampled at weekly intervals in 1984, and acidic waters were found to be widespread. Of the 150 stream sampling points (sampled weekly) 13% had a mean pH of less than 5.5 and 34% had a minimum pH of less than 4.5. Analyses for "dissolved" (filterable at 0.45 µm) aluminium showed mean values of $\geq 0.1$ mg/L in 28% of streams, and maximum values of $\geq 0.2$ mg/L in 48% of streams.

Adverse ecological effects have been noted in streams and lakes with low pH and high aluminium concentrations. Previously productive salmonid lake fisheries in Mid Wales have become extinct where such conditions have developed. Field experiments during acidic, high-aluminium stream episodes resulted in mortalities of caged salmonids (Gee and Stoner, in press), and artificially elevated aluminium levels caused increased invertebrate drift (Ormerod *et al.*, 1987).

Some private water supplies in upland areas may contain elevated aluminium levels deriving from natural sources where no pH correction is carried out. However the most likely reason for the presence of aluminium in public supplies is the use of aluminium sulphate as a coagulating agent in the water treatment process. This is currently the most effective method for treating highly coloured, low-conductivity waters.

The maximum admissable concentration (MAC) for aluminium in the EC Drinking Water Directive (80/778/EEC) is 0.2 mg/L, applying to water at the consumer's tap. Where a supply fails to comply with the MAC, and public health is not prejudiced, water authorities may apply for derogations or delays. These are related respectively to geological conditions and treatment inadequacies (or corroded mains). The WWA currently has derogations on three small treatment works and these will all be abandoned by 1989. There are delays on 18 treatment works, with compliance dates of not later than 1990.

## References

Donald, A.P. and Stoner, J.H. The quality of atmospheric deposition in Wales. *Arch. Environ. Contam. Toxicol.*, (in press).
Gee, A.S. and Stoner, J.H. The effect of seasonal and episodic variations in water quality on the ecology of upland waters in Wales. *Environ. Pollut.*, (in press).
Ormerod, S.J., Boole, P., McCahon, C.P., Weatherley, N.S. and Edwards, R.W. (1987). Short-term experimental acidification of a Welsh stream: comparing the biological effects of hydrogen ions and aluminium. *Freshwater Biology*, 17, 341-356.
Stoner, J.H., Gee, A.S. and Wade, K.R. (1984). The effects of acidification on the ecology of streams in the upper Tywi catchment in West Wales. *Environ. Pollut.*, 35, 125-157.

# 23 The Solution Chemistry of Aluminium and Silicon and Its Biological Significance

J.D. Birchall and J.S. Chappell
*Imperial Chemical Industries PLC, P O Box 8, The Heath, Runcorn, Cheshire WA7 4QE, England*

## Summary

*Aluminium and silicon can enter the aquatic environment by leaching from the aluminosilicates of rocks and soil minerals. Aluminium can exert toxic effects whereas silicon is claimed as an essential element. It is shown that the two elements interact in aqueous solution to form soluble aluminosilicate species, even in the presence of organic complexing agents, so reducing the availability and toxicity of aluminium and possibly accounting for some aspects of the essentiality of silicon.*

## Introduction

After oxygen, the two most abundant elements in the earth's crust are silicon and aluminium (21 and 6 atomic percent respectively) combined as the aluminosilicates of rocks and soil minerals. Both elements are leached into the aquatic environment (with the leaching of aluminium being favoured by acidic conditions) and both can then interact with biology. Whereas aluminium is increasingly regarded as being capable of exerting toxic activity (*e.g.* to fish), silicon is listed as an essential element. The silicon cycle is shown in Figure 1 (Birchall, 1978) in which is also indicated the preferential release of aluminium from aluminosilicates at low pH (Figure 2). The mechanisms by which silicon exerts its reported beneficial effects and those by which aluminium exerts toxic effects are not clear, but the two oxides have a peculiar affinity in solid state chemistry, colloid chemistry and in aqueous solution (Iler, 1979). What is known of both elements in biology is very briefly reviewed below.

### Silicon

Silicon was listed as an essential element following studies of the effects of silicon deficiency in animals (Carlisle, 1972; Schwarz, 1972). In rats and chicks, silicon

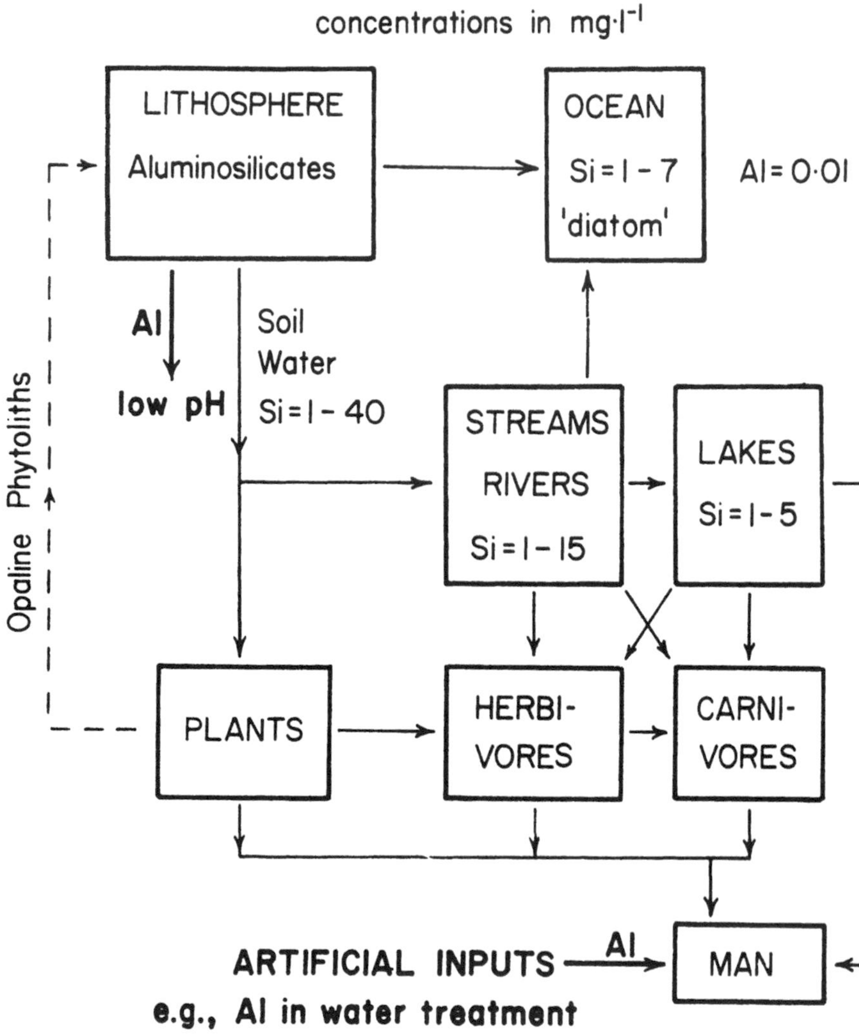

concentrations in mg·l⁻¹

Figure 1. *The silicon cycle.*

deficiency produced low weight gains - reversed by silicon supplementation - and profound changes in osteogenesis associated with a failure to properly synthesise the bone organic matrix. A major effect of silicon deficiency appears to be on collagen synthesis, not only in bone organic matrix, but generally, and there is also a marked reduction in glycosaminoglycan synthesis.

Silicon is found at the mineralisation front in bone (Carlisle, 1986a). In an attempt to penetrate to the molecular basis of these findings, prolyl hydroxylase from Si-deficient sources was shown to have its low activity very significantly increased by the addition of silicic acid. It has long been known that silicon is an essential element for the diatom, not only for the construction of the silicious skeleton or

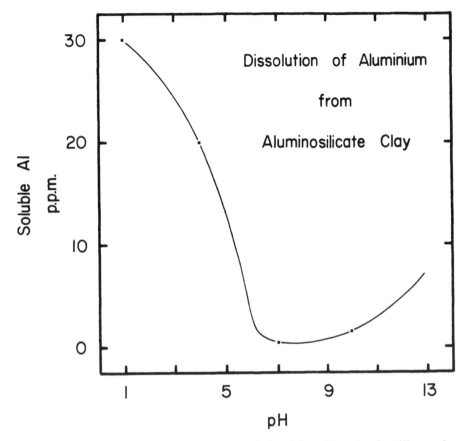

**Figure 2.** *Effect of pH on the dissolution of aluminium from aluminosilicate clay.*

frustule but also for the maintenance of major metabolic processes (Werner, 1977). Certain plants too, notably rice, are regarded by some workers as having a requirement for silicon.

A major problem in seeking to understand the molecular basis for the biological effects of silicon is that no organic binding of silicon has been convincingly demonstrated in biology, the esters and complexes of silicic acid, $Si(OH)_4$, the form in which silicon enters biology, being unstable in aqueous solution at pH 7.4. The role of silicon remains an enigma. The reader is referred to monographs such as the Ciba Foundation Symposium No. 121, 1986 on "Silicon Biochemistry" and to Bendz and Lindqvist (1978).

*Aluminium*
Aluminium is now known to be associated with toxic effects in higher animals. The incidence of serious disorders in patients on long-term dialysis showed a geographic distribution related to the concentration of aluminium in the water used (Platt *et al.*,

**Table 1.** *The effect of silicic acid on the inhibition of prolyl hydroxylase activity by aluminium (Birchall and Espie, 1986).*

| Addition - all co-factors plus (100 μM) | Percentage inhibition over control |
|---|---|
| Si(OH)$_4$ | 0 |
| Fe then Al | 20 |
| Al then Fe | 55 |
| Al + 6 Si(OH)$_4$ | 0 |

1977). Three disorders could result: dialysis encephalopathy, dialysis osteomalacia and a severe anaemia unresponsive to iron therapy, but reversible when low Al dialysis fluid is used. In bones with dialysis osteomalacia aluminium has been found to be located at the cement line (McClure, 1984). An early demonstration of the neurotoxicity of aluminium was by exposing rabbits to aluminium salts (Klatzo *et al.*, 1965), and the obvious neurofibrillary degeneration observed prompted an examination of the brains of patients having Alzheimer's disease (Crapper-McLachlan *et al.*, 1973) and the claim that they contained higher than normal aluminium levels. The people of the island of Guam have a high incidence of neurodegenerative disease, in particular a form of Parkinsonism with severe dementia and amyotrophic lateral sclerosis with widespread neurofibrillary tangle formation. The neurons bearing tangles have been found to contain aluminium (Perl, 1985). This finding, with others, has prompted questions as to the relationship between Alzheimer's disease and exposure to aluminium (Perl, 1985) recently catalysed by the finding of aluminosilicates within the cores of plaques in Alzheimer brains (Candy *et al.*, 1986).

Aluminium is also associated with the toxicity to fish of acidified waters (Karlsson-Norrgren *et al.*, 1986) and, generally, evidence is growing that biology is vulnerable to available aluminium. This paper addresses the issue of the role of silicon in determining the bio-availability of aluminium, a hitherto neglected consideration.

## The Potential Importance of the Interaction of Aluminium and Silicon

In considering the mechanism by which silicon (as silicic acid) may influence biochemical processes, Birchall and Espie (1986) concluded that the only likely physiological reactions of silicic acid were with metal ions, reactions with organic species being unlikely in aqueous systems at physiological pH. These authors pointed to the unique affinity between Si(OH)$_4$ and aluminium, for example, as the aluminate ion, Al(OH)$_4^-$, a dominant species at physiological pH. In an investigation of the activating effect of silicic acid on the activity of prolyl hydroxylase obtained

234

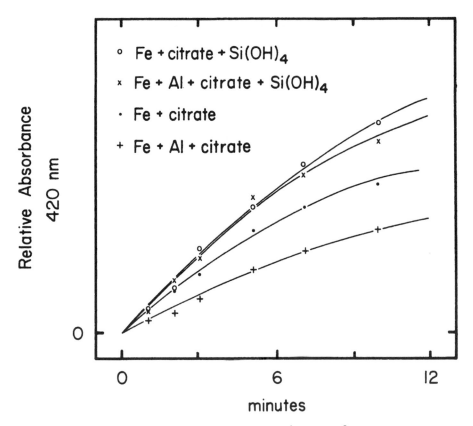

**Figure 3.** *Competition of Fe, Al (each at $2.0x10^{-4}M$) for $M^{3+}$ binding to desferrioxamine ($1.7x10^{-4}M$). All curves include citrate ($1.0x10^{-3}M$). Curves a (•) and b (+) are with citrate alone, curves c (o) and d (x) include silicic acid ($1.0x10^{-3}M$).*

from Si-deficient sources (Carlisle, 1986a) they hypothesised that aluminium contained within the system could deactivate the enzyme by replacing the iron which is necessary for its function. It was shown (Table 1) that the introduction of aluminium reduced the *in vitro* activity of this enzyme, especially when presented *before* iron and that this effect of aluminium was completely prevented when silicic acid was present.

It was concluded that silicic acid selectively interacted with aluminium, removing it from competing with iron for binding in the enzyme. This is an interesting finding since it suggests that under normal physiological conditions any aluminium present is made unavailable in the presence of silicic acid, perhaps by the formation of aluminosilicate species in solution or as colloids. The nature of the interaction of aluminium and silicon in aqueous solution may thus be of general importance in biology.

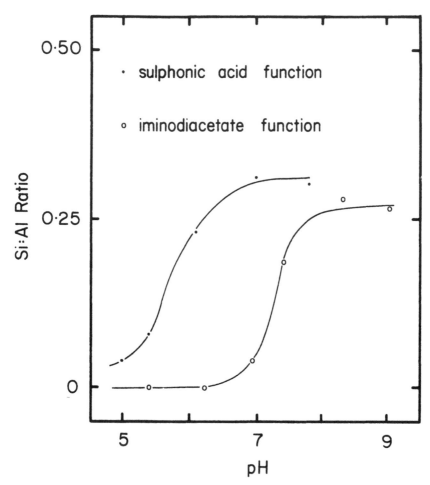

**Figure 4.** *Si:Al ratio of the retained molecular species as a function of solution pH. Species are retained by an ion exchange resin with either sulphonic acid (•) or iminodiacetate (o) functional groups from solutions of aluminium (1.0x10⁻⁴M) and silicic acid (5.0x10⁻⁴M).*

**The Interaction of Silicic Acid with Dissolved Aluminium Species**

*Effect of silicic acid on the competitive binding of iron and aluminium by desferrioxamine*

In an attempt to model the finding with prolyl hydroxylase, the binding of $Fe^{3+}$ and aluminium by the strong $M^{3+}$ chelator, desferrioxamine, was studied with and without the addition of silicic acid. The pH was 7.4 and citrate was used to prevent the rapid precipitation of metal hydroxides. The binding of $Fe^{3+}$ was followed by measuring absorbance at 420 nm. The results are shown in Figure 3 from which a number of conclusions are possible. Firstly, from the reduced absorbance (curves a and b) it is clear that aluminium competes for binding with $Fe^{3+}$ in the early stages of the reaction. The addition of silicic acid to the system resulted in two effects. The rate of binding of $Fe^{3+}$ was increased when silicic acid was present, so that citrate and silicate co-operate in presenting iron to desferrioxamine by retarding polymerisation. However, the competition of aluminium was reduced (curves c and d). Thus, it appears that, even in the presence of citrate, silicic acid interacts with aluminium selectively and sufficiently strongly for a marked effect on its binding to desferrioxamine to be observed in the early stages of the reaction.

**The formation of aluminosilicate species**

Solutions of aluminium ($1 \times 10^{-4}$M) with silicic acid ($5 \times 10^{-4}$M) were aged at various values of pH and passed through filters retaining particles > 0.22 μm. The retained solids were recovered by dissolution in 3N HCl and the extracts analysed by atomic absorption spectroscopy. Similar solutions, aged 24 hours, were passed through columns containing ion exchange resins (Amberlite) with (a) sulphonic acid and (b) iminodiacetate functional groups. Retained species were removed from the resins by extraction with 3N HCl and analysed.

The solids isolated by filtration typically required 3 to 5 weeks to form under the low ionic strength conditions (0.01M NaCl) employed.

The Si:Al ratio of these colloidal precipitates fell in the relatively narrow range of 0.3 - 0.6 for all solutions in the pH range of 5 - 9; however, for solutions aged 10 weeks at pH 8 the Si:Al ratio tended to rise above 0.6 but remained less than 1.0. These findings are qualitatively consistent with those of Farmer *et al.* (1979) who report that aluminosilicate species with Si:Al ratio of 0.25 - 0.5 can form in aqueous solution at pH ≤ 5.5. These species appear to be precursors of amorphous or poorly crystalline aluminosilicate solids and may be considered as fragments of the structure of the mineral imogolite for which the ideal composition is $(HO)_3Al_2O_3SiOH$, *i.e.* Si:Al ratio of 0.5.

The molecular species which give rise to the solid phase were detected by the ion exchange technique in which species from solutions containing aluminium and silicic acid were retained by the resins. Figure 4 shows the average Si:Al ratio of the retained species as a function of pH for both types of resin. The molecular species throughout intermediate pH apparently develop into the same solid phase with the composition seen in the filtration results. The Si:Al ratio from the ion exchange work is observed to be a strong function of pH with the maximum level of the ratio (0.25 - 0.35) in agreement with the composition range of the solid phase. This behaviour apparently reflects the relative stability of the aluminosilicate species in the presence of the functional groups of the resins. There is no evidence that the *rate* of formation for the molecular species is significantly variable with pH.

The sulphonic functional group binds the species exclusively by ionic (electrostatic) interaction and thus solely retains the cationic species in a loose association. Stable aluminosilicate species are indicated for pH ≥ 6, although their stability is low at pH < 5. The iminodiacetate functional group is a much stronger binder, forming tight coordination bonds directly with the aluminium atoms of the

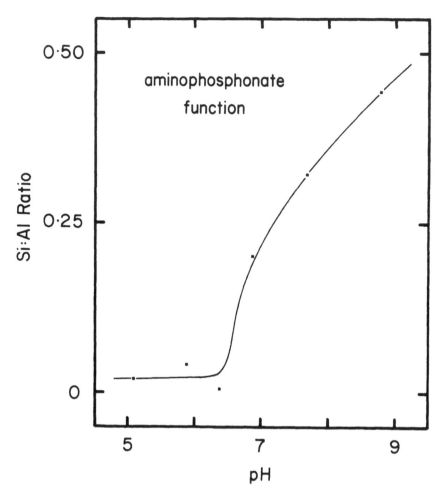

**Figure 5.** *Effect of pH on the Si:Al ratio of the species retained by aminophosphonate functional groups.*

aluminosilicate species. The aluminium-silicate bonds remain stable at pH > 7. The silicate ligand is readily displaced by the iminodiacetate groups at pH < 7. This transition in stability as a function of pH correlates with a change-over in the coordination of aluminium from cationic, octahedral complexes (pH < 6) to the anionic, tetrahedrally-coordinated aluminate $Al(OH)_4^-$ (at pH > 7). This change in the coordination geometry may carry over to the structure of the aluminosilicate molecular species with a concomitant enhancement in stability at pH > 7. It is concluded that molecular species of aluminosilicate form in the presence of organic complexing agents of aluminium (*e.g.* citrate) but these agents prevent aggregation to form colloidal solid phases such as imogolite. Solutions of aluminium ($1 \times 10^{-4}$M) and equimolar amounts of silicic and citric acids ($5 \times 10^{-4}$M) at physiological pH yield no precipitate after 12 weeks, although 1 day-old solutions passed over the ion exchange resins revealed species with Si:Al ratios $\geq 0.5$.

Silicic acid appears then to play an important role in limiting the bio-availability of aluminium in aquatic environments. It is, for example, known that the equilibrium solubility of aluminium in sea water at pH 8.1 is limited by the presence of silicic acid (Willey, 1975). The balance of the two elements, in conjunction with pH, may be of considerable environmental consequence, for example, in determining the response of fish to aluminium.

*Biological binding sites for aluminium and the effect of silicic acid on binding*

Aluminium is known to interact with membranes (Vierstra and Haug, 1978), DNA (Karlik *et al.*, 1980) and to inhibit hexokinase (Womack *et al.*, 1979). The activity of the latter is reduced as a result of the binding of aluminium to ATP in place of $Mg^{2+}$ and activity is restored by the addition, for example, of citrate. The affinity of aluminium for ATP is nearly 7 orders of magnitude greater than that of $Mg^{2+}$ (Martin, 1986) and it has been suggested that its binding to ATP stabilises the terminal phosphoryl group against transfer to glucose by hexokinase (Trapp, 1980). The complex of $Al^{3+}$ with citrate is approximately 16 times more stable than ATP-Al (Martin, 1986), accounting for the ability of citrate to activate the system (Viola *et al.*, 1980).

These various findings indicate that a major binding site for aluminium in biology is phosphate, suggesting that the competition between phosphate and silicic acid for binding aluminium may be of particular significance. The ability of citrate to remove $Al^{3+}$ from ATP, and the fact that aluminosilicates can form in the presence of citrate, points to a possible important role for silicic acid.

The precipitate formed from solutions containing $1 \times 10^{-4}$M Al and equimolar concentrations ($5 \times 10^{-4}$M) of phosphate and silicate varies in composition depending upon pH. At pH 7.4 the solid contained an Si:Al ratio of 0.44 whereas at pH 6.4 the silicon content was negligible with an Si:Al ratio of 0.02. This is consistent with the finding that, in soil science, the adsorption of silicate and phosphate on alumina-rich soils is pH-dependent, with silicic acid adsorbing in preference to phosphate (and displacing the latter) at pH greater than about 6.6 (Obihara and Russell, 1972).

To confirm the pH-dependent interaction between aluminium, phosphate and silicic acid, solutions containing aluminium ($1.0 \times 10^{-4}$) and silicic acid ($5 \times 10^{-4}$M) were passed over an ion exchange resin with aminophosphonate functional groups. The composition of the retained species reflects the stability of the aluminosilicate entity in the presence of phosphonate groups. The Si:Al ratio of the retained species as a function of solution pH is shown in Figure 5. The formation of aluminosilicates is favoured above pH 6.6 whilst below this pH aluminium bonds preferentially to phosphonate. Binding is seen to be delicately balanced around physiological pH, possibly with binding to silicate at plasma pH and to phosphate at intracellular pH.

**Discussion**

The fact that aluminosilicate solution species (and eventually colloids and amorphous solid phases) are formed when aluminium and silicic acid interact in aqueous solutions at pH ≥ 5 suggests an important role for silicon in restricting the bio- availability of aluminium and reducing its toxicity in an aquatic environment. This role for silicon does not appear to have received consideration. Aluminosilicate species form in solution even in the presence of organic compounds capable of complexing aluminium (*e.g.* citrate) although the presence of citrate, *etc.* appears to retard or prevent actual precipitation. Whilst the ameliorating effect of organic compounds (*e.g.* humic acids) on aluminium toxicity has been considered, it appears likely that the effect of silicic acid may be of even greater and more general importance. It is possible that the only other common anion with as marked an effect is F⁻ but this, unlike silicic acid, is itself toxic at relatively low levels.

The ameliorating effect of silicic acid on the toxicity of aluminium to fish is the subject of a current investigation (Exley, 1987) with early results tending to confirm the effect. More general issues are raised by the interactions between aluminium and silicic acid demonstrated here. Silicon is regarded as an essential element for some species and it is a normal component of plasma and tissue with an average level in human serum of about 50 μg/dL (Carlisle, 1986b). It is possible that silicic acid restricts the adsorption of aluminium into the gut and the interaction of silicic acid with aluminium in plasma and at the cellular level must be of interest should aluminium gain entry. Aluminium is found in diseased neurons and Alzheimer senile plaques contain aluminosilicates. Aluminium occurs also in cardiovascular tissue (Webb *et al.*, 1974) as does silicon (Carlisle, 1974).

It is known that aluminium interacts with ATP, membranes, DNA, *etc.* via phosphate groups and the present study reveals that this interaction may be inhibited by silicic acid at physiological pH. Interaction via phosphate with membranes and within the ATP-rich chloride cells of fish gills seems a possible mechanism for the toxic action of aluminium (Karlsson-Norrgren *et al.*, 1986). From the results presented here this would be expected to be ameliorated by silicic acid.

As regards the toxicity of aluminium in higher animals, a generalised binding to phosphate seems too non-specific but it is worth noting that silicic acid will affect this binding only at pH 6.6 and thus may be ineffective at intra-cellular pH. It is also conceivable that aluminium binding is strong to specific minority phosphate - containing entities, for example, where two or more phosphate groups could tightly chelate as in the inositol triphosphates *e.g.* I - 1,4,5-$P_3$.

The pathological effects of Si-deprivation have not been mechanistically explained but it is conceivable that the removal of silicon upsets a delicate balance between silicon and aluminium and unleashes the activity of aluminium.

It is interesting to reassess the "water factor" in the variation of the geographical incidence of IHD and stroke and in which an inverse correlation with water hardness is found (Masironi, 1979). Silicon tends to be high in "mineralised" hard waters and a negative correlation with IHD/stroke incidence is reported for silicon (Pocock *et al.*, 1986). Aluminium seems not to have been considered but since, of the common metallic aquatic contaminants, silicon has a chemistry only with aluminium it would seem sensible to explore the balance of these two elements as a factor. In natural waters complexes are not formed between $Ca^{2+}$, $Mg^{2+}$ and silicic acid (Iler, 1979).

It is an intriguing possibility that evolving biology has avoided aluminium by a pH sufficiently high to avoid excessive dissolution and/or with an excess of silicic acid able to reduce the availability of aluminium at pH values above about 5.5.

## Acknowledgement

The authors wish to thank W. Correia of the Central Analytical Facilities, MIT for his work on atomic absorption analysis.

## References

Bendz, G. and Lindqvist, I. (eds.) (1978). *Biochemistry of Silicon and Related Problems*. Plenum Press, New York.

Birchall, J.D. (1978). Silicon in the biosphere. In: R.J.P. Williams and J.R.R.F. Da Silva (eds.), *New Trends in Bio- Inorganic Chemistry*, pp.209-252. Academic Press, London.

Birchall, J.D. and Espie, A.W. (1986). Biological implications of the interaction (via silanol groups) of silicon with metal ions. In: *Silicon Biochemistry*, Ciba Foundation Symposium No.121, pp.140-153. John Wiley, Chichester.

Candy, J.M., Oakley, A.E., Klinowski, J., *et al.* (1986). *Lancet*, 1, 354-357.

Carlisle, E.M. (1972). *Science*, 178, 619-621.

Carlisle, E.M. (1974). *Fed. Proc.*, 33, 1758-1766.

Carlisle, E.M. (1986a). Silicon. In: W. Mertz (ed.), *Trace Elements in Human and Animal Nutrition*, pp.128-133. Academic Press, Orlando, Florida.

Carlisle, E.M. (1986b). Silicon as an essential trace element in animal nutrition. In: *Silicon Biochemistry*, Ciba Foundation Symposium No.121, pp.123-139. John Wiley, Chichester.

Crapper, D.R., Krishnan, S.S. and Dalton, A.J. (1973). *Science*, 180, 511-513.

Exley, C. (1987). Personal Communication, University of Stirling.

Farmer, V.C., Fraser, A.R. and Tait, J.M. (1979). *Geochim. Cosmochim Acta*, 43, 1417-1420.

Iler, R.K. (1979). *The Chemistry of Silica*. Wiley-Interscience, New York.

Karlik, S.J., Eichhorn, G.L., Lewis, P.N. and Crapper, D.R. (1980). *Biochemistry*, 19, 5991-5998.

Karlsson-Norrgren, L., Bjorklund, I.B., Ljungberg, O. and Runn, P. (1986). *J. Fish Dis.*, 9, 11-25.

Klatzo, I., Wisniewski, H. and Streicher, E. (1965). *J. Neuropath. Exp. Neurol.*, 24, 187-199.

Martin, R.B. (1986). *Clin. Chem.*, 32, 1797-1806.

McClure, J. and Smith, P.S. (1984). *J. Pathol.*, 12, 293-299.

Masironi, R. (1979). *Phil Trans. R. Soc. Lond.*, B288, 193-203.

Obihara, C.H. and Russell, E.W. (1972). *J. Soil Science*, 23, 105- 17.

Perl, D.P. (1985). *Environ. Health Persp.*, **63**, 149-153.

Platts, M.M., Goode, G.C. and Hislop, J.S. (1977). *Brit. Med. J.*, **10**, 657-660.

Pocock, S.J., Shaper, A.G., Powell, P. and Packham, R.J. (1986). The British regional heart study: cardiovascular disease and water quality. In: I. Thornton (ed.),*Proceedings of the First International Symposium on Geochemistry and Health*, pp.141-157. Science Reviews, Northwood, Middlesex, UK.

Schwarz, K., and Milne, D.B. (1972). *Nature*, **239**, 333-334.

Trapp, G.A. (1980). *Neurotoxicology*, **1**, 89-100.

Vierstra, R. and Haug, A. (1978). *Biochem. Biophys. Res. Comm.*, **84**, 138-143.

Viola, R.E., Morrison, J.F. and Cleland, W.W. (1980). *Biochemistry*, **19**, 3131-3137.

Webb, J., Kirk, K.A., *et al.* (1974). *J. Mol. Cell. Cardio.*, **6**, 383-394.

Werner, D. (1977). Silicate metabolism. In: D. Werner (ed.), *The Biology of Diatoms*, pp.110-149, Blackwell Scientific, Oxford.

Willey, J.D. (1975). *Mar. Chem.*, **3**, 227-240.

Womack, F.C. and Colowick, S.P. (1979). *Proc. Nat. Acad. Sci. (USA)*, **16**, 5080-5084.

# 24 Human Exposure to Environmental Aluminium

K.C. Jones
*Department of Environmental Science, University of Lancaster,*
*Bailrigg, Lancaster LA1 4YQ, England*

## Summary

*Representative values of Al in the environment and in man have previously been selected from available data and a pathway analysis performed utilising the exposure commitment method (Jones and Bennett, 1985; 1986).*

*Using a derived estimate of the body burden (60 mg), a representative value for dietary intake (20 mg/day) and a fractional absorption of 0.01, a mean retention time of Al in the body of 300 days is obtained. This corresponds to a biological half-life of 210 days. The assessment indicates that an average dietary intake rate of 20 mg/day contributes 660 µg Al/kg body weight, while inhalation of Al in air makes, in comparison, a negligible contribution to the body content.*

## Introduction

Recent literature indicates a considerable interest in the role of aluminium in the "acid rain" problem, and in its neurotoxicological properties. We have previously been concerned with the pathways of human exposure to aluminium utilising the exposure commitment method. Exposure commitments give a basis for comparing contributions to pollutant exposure from various pathways and for estimating equilibrium concentrations resulting from continuing releases. The method has been developed at the Monitoring and Assessment Research Centre (MARC) for GEMS, the Global Environmental Monitoring System and is being applied to the assessment of pollutant transport in the regional and global environment. The objectives of the method are to determine the partitioning of pollutant amounts in pathway movements and the amounts which ultimately reach the receptor.

## General Comments

Aluminium is the third most abundant element in the Earth's crust, yet is present only in very small amounts in human tissues. This is because Al is largely excluded from

This work was undertaken at the Monitoring and Assessment Research Centre, King's College London, University of London, 459A Fulham Road, London SW10 0QX in collaboration with Dr B.G. Bennett.

**Table 1.** *Current levels of aluminium in the background environment and in man.*

|  | Air | Soil | Diet | Bone | Body |
|---|---|---|---|---|---|
| **Inhalation pathway** | | | | | |
| - Urban | $1,000$ ng/m$^3$ | | | 2.6 µg/kg | |
| | | | | | 0.5 µg/kg |
| - Rural | $200$ ng/m$^3$ | | | 0.5 µg/kg | |
| | | | | | 0.1 µg/kg |
| **Ingestion pathway** | $200$ ng/m$^3$ | 70 µg/g | 20 mg/day | 3,400 µg/kg | |
| | | | | | 600 µg/kg |
| Total | | | | ~3,400 µg/kg | ~600 µg/kg |

the body by very inefficient absorption via the gastro-intestinal tract. However, abnormally high concentrations have been reported in individuals undergoing renal dialysis treatment, where exposure is via the dialysate prepared from tap water. Increased amounts of the Al have also been reported in the brain of subjects suffering from Alzheimer's disease, although it is not yet known whether Al is the causative agent.

**Pathway Analysis**

The transfer of Al to man from general environmental sources occurs via the inhalation and ingestion pathways. The procedure of the pathways analysis has been described elsewhere (Bennett, 1981) and the reader is referred to the fuller papers for the detailed assessment. Table 1 summarises the current levels of Al in the background environment and in man.

**References**

Bennett, B.G. (1981). Exposure commitment concepts and application, In: *Exposure commitment assessments of environmental pollutants* Vol.1, No.1, MARC Report. No.23. Monitoring and Assessment Research Centre, King's College, University of London.

Jones, K.C. and Bennett, B.G. (1985). *Exposure commitment assessments of environmental pollutants* Vol.4. Aluminium MARC Report No.33. Monitoring and Assessment Research Centre, King's College, University of London.

Jones, K.C. and Bennett, B.G. (1986). Exposure of man to Environmental aluminium - an exposure commitment assessment. *Science of the Total Environment* **52**, 65-82.

# 25 Geographical Associations between Aluminium in Drinking Water and Registered Death Rates with Dementia (including Alzheimer's Disease) in Norway

Trond Peder Flaten
*Department of Chemistry, College of Arts and Science, University of Trondheim, N-7055 Dragvoll, Norway*

## Summary

*Comparisons of maps and correlation and regression analysis indicate a geographical association between aluminium in drinking water and dementia in Norway. A major uncertainty, however, relates to the use of registered death rates with dementia (cause mentioned on the death certificate) as a measure of incidence rates of Alzheimer's disease. It is possible that the association might be due to differences in diagnosis and registration of dementia.*

## Introduction

Aluminium (Al) is present in neurofibrillary tangle-bearing neurons (Perl *et al.*, 1980; 1986; Crapper McLachlan and De Boni, 1980), and apparently also in the cores of neuritic (senile) plaques (Candy *et al.*, 1986), the two neuropathological hallmarks of Alzheimer's disease (AD). Other pieces of evidence (see *e.g.* Ganrot, 1986; Flaten, 1986) also indicate that Al somehow is related to AD. It is, however, impossible to state at present whether Al is involved in the *aetiology* of AD, or is concentrated secondarily as an *effect* of the disease. Al exists in the environment predominantly in forms of low availability to man and most other biological species (Driscoll *et al.*, 1984; Ganrot, 1986). However, *acid precipitation* has in recent years led to a leaching of Al from the soils in acid rain-sensitive areas, leading to elevated levels of Al in drinking water (see below). The possibility also exists that some local foodstuffs in these areas may contain elevated Al levels, *e.g.* in the form of certain *chelates* in plants. Some concern has been raised that these possible acid rain-induced increments in bioavailability of Al may constitute a threat to human health (Lancet, 1985; Nordberg *et al.*, 1985; Vogt, 1986). In Norway, different

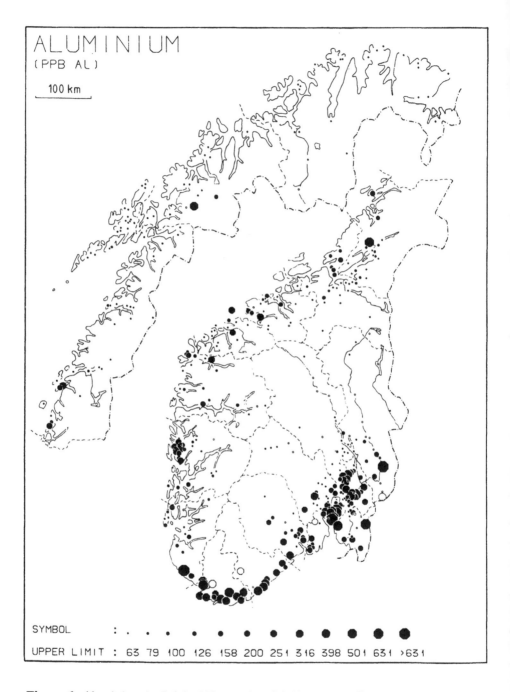

**Figure 1**. *Aluminium in finished Norwegian drinking water. Concentration limits given in ppb (μg Al/L). Waterworks using surface water are marked with filled symbols; waterworks using ground water are marked with open symbols.*

regions are affected by acid precipitation to a very varying degree (*cf.* Figure 1 and Wright *et al.*, 1976). Therefore, it seemed interesting to try to investigate whether AD rates were elevated in the parts of Norway most affected by acid precipitation. Data on the incidence of AD are not available on a regional level like the Norwegian municipality, so *registered death rates with dementia* have been used in the present study as an approximate measure of AD rates. AD is the most common disease leading to the clinical picture of dementia, and is generally considered to constitute more than half of the total number of dementia cases (Tomlinson *et al.*, 1970; Katzman, 1986).

**Materials and Methods**

Four water samples, one for each season, were collected in the period October 1982 to August 1983 from each of 384 Norwegian waterworks that supply 70.9% (2.91 million persons) of the total population in Norway. Of the waterworks, 349 are based on surface water (lakes or rivers), the remaining 35 use groundwater. On each occasion, cleaned (Laxen and Harrison, 1981, cleaning method "D") polythene bottles containing 2 mL 1:4 $HNO_3$ (Suprapur, Merck) were sent to the waterworks. The personnel at the waterworks were asked to collect samples after water treatment *i.e.* from taps at the waterworks, after water has been run through the taps for at least five minutes. The collected samples were returned by mail and analysed in four batches (random order) by ICP spectrometry. Various control samples (synthetical standards, duplicates, blanks and previously analysed samples) were analysed at random intervals together with the regular samples.

Mortality data (ICD 8) were provided by the Central Bureau of Statistics of Norway and made available through the research group "Medical Geography in Planning" in Trondheim. Dementia is generally not considered to be a disease of high fatality. For the 10 year period 1974-83, dementia was coded as the underlying cause of death from 543 (males) and 990 (females) death certificates for Norway as a whole. However, in Norway, contrary to the situation in most other countries, up to three *contributory* causes-of-death are coded per death certificate in addition to the underlying cause. For the same 10 year period, dementia was coded as a contributory cause from 5,099 (males) and 8,095 (females) certificates. Age-adjusted death rates were therefore calculated on the basis of the total number of death certificates from which dementia (senile or presenile dementia, ICD 8 290.0 and 290.1, respectively) had been coded *either* as the underlying *or* as a contributory cause of death. Since many of the Norwegian municipalities have a very small population, geographically neighbouring municipalities were pooled to aggregates of 10,000+ inhabitants, and death rates were calculated for the resulting 193 geographical units. Only six (males) and two (females) of these units have less than 10 registered dementia cases in this 10 year period. Hence, in most of the units the number of cases is considerable.

In many of the 193 municipality aggregates, Al levels in the drinking water of major parts of the population have not been analysed. However, acidification in Norway is a phenomenon occurring on a large geographical scale, and in most cases the Al concentrations of neighbouring lakes and streams are fairly similar (see Figure

1). Therefore, for those aggregates where the Al data were incomplete, Al levels were estimated from neighbouring values, excluding waterworks adding aluminium sulphate (see below) and waterworks where high Al values probably were due to particulate matter (*cf.* Flaten, 1986). In the estimation procedure, Al concentrations of 155 small lakes in South Norway sampled in 1974 (Wright *et al.*, 1976) were also used.

For the 193 municipality aggregates, correlation and regression analysis was performed between the Al values and the dementia rates. Since registered death rates with dementia seem to be higher in *urban* than in *rural* municipalities (Vogt, 1986, p.55), the percentage of the population living in "densely populated areas" in 1970 (CBS, 1974) were calculated for the 193 aggregates, and correlation analysis was performed between this variable and registered death rates with dementia. A stepwise regression analysis where Al was added to the model *after* the percentage living in densely populated areas was also performed.

## Results and Discussion

The analytical precision (calculated from the results for duplicates) was better than 6% above 0.1 mg Al/L, and increased gradually below this concentration to 30% at 0.02 mg Al/L. The accuracy (calculated from the results for standard solutions) was better than 10% above 0.1 mg Al/L, and increased gradually below this concentration to 30% at 0.04 mg Al/L. The highest analytical value found among 20 blank solutions was 0.018 mg Al/L.

Concentrations of Al ranged from not detectable (10% of the samples were found to have Al concentrations <0.008 mg/L) to 4.10 mg/L (in waterworks adding aluminium sulphate). The median was 0.055 mg/L, and the 90 percentile was 0.238 mg/L. The median compares favourably with results from a recent survey of drinking water in the USA, where 55% of the raw surface waters had Al concentrations 0.05 mg/L (Miller *et al.*, 1984).

The concentration of aluminium (Figure 1) is comparatively high (typically 0.15 - 0.4 mg/L) along the south and southeastern coasts, in some areas in the inland of southeastern Norway and in Bergen on the west coast. This is most likely due to acid precipitation in these parts of the country, causing mobilisation of Al in the soil (Wright *et al.*, 1976). In North Norway and in the inland parts of South Norway drinking water very seldom contains more than 0.05 mg Al/L.

Aluminium sulphate is added as a coagulating agent to remove humic substances and particulate matter at 17 (4.4%) of the waterworks. Water treatment efficiency varies from one treatment plant to another: mean Al concentrations in treated water range from 0.03 to 3.04 mg/L. The number of these waterworks is too small to have any marked influence on the geographical pattern. Most Norwegian waterworks use a very mild degree of water treatment, so the Al values in the treated drinking water samples mostly reflect the values in the water sources.

The seasonal variability in the Al levels is generally about 20- 40% (Flaten, 1986). The mean values for the four samples taken at different seasons are, in most cases, probably fairly accurate estimates of the actual drinking water intakes of Al.

For the municipalities where Al data were incomplete or lacking, and had to be estimated, the uncertainties are of course larger, but probably not of vital importance, because acidification in Norway is a large-scale phenomenon. Therefore, Al concentrations in neighbouring lakes and streams in most cases are fairly similar.

Drinking water normally contributes little to the *total intake* of Al. At high concentrations (0.2-0.4 mg/L), as in southern Norway, intake through drinking water is seldom more than 0.5 mg/day. The total human daily intake of "natural" Al (principally through foodstuffs) is probably some 2-10 mg/day (Koivistoinen, 1980; Greger, 1985). In addition, similar or somewhat higher amounts may be ingested as *food additives* (Lione, 1983; Greger, 1985). Much higher amounts may be ingested in the form of certain *medications*, especially some antacids (Lione, 1983). This intake may amount to 1,000 mg/day or more. These considerations of Al intake would seem to imply that there is little need for concern about the small amounts of Al ingested through drinking water. However, since acid water probably releases more aluminium from *cookware* than does neutral water, some foods prepared in aluminium cookware with acid water may possibly contain more aluminium than the same food prepared with neutral water. Potentially more important, little is known about the *bioavailability* (or "neuroavailability") of different Al species. Generally, Al is poorly absorbed through the intestine (Ganrot, 1986), and some Al species found in *e.g.* juice or drinking water may be more easily absorbed than others. Moreover, the possibility exists that some local foodstuffs in acid rain- sensitive areas may contain elevated levels of Al, *e.g.* in the form of certain *chelates* in plants. Such chelates may be more biologically available than other Al species. For example, it has been shown that Al-*citrate* is much more readily absorbed from the intestinal tract than Al-*hydroxide*, both in rats and humans (Slanina *et al.*, 1984; 1985; 1986; Weberg and Berstad, 1986). Furthermore, a complex between Al and maltol has recently been described by Finnegan *et al.*, (1986). This complex is of neutral charge in aqueous solutions and remarkably stable to hydrolysis, and it seems to be an unusually potent neurotoxin.

A choropleth map for death rates with dementia for females is given in Figure 2. The map for males (not shown here) is very similar. These maps bear some resemblance to that of Al in drinking water (Figure 1); the high-aluminium area along the south and southeastern coasts is fairly clearly reflected in high rates on the dementia maps, and North Norway and most of the inland areas of South Norway generally have low dementia rates and low levels of Al in drinking water. When interpreting the map, it should be noted that the population density varies greatly between different parts of Norway. In general, the larger the area of a municipality aggregate, the more sparsely populated. For example, the high-dementia area across the inland of South Norway is sparsely populated.

Scatter plots, correlation coefficients and regression lines for the 194 municipality aggregates are given in Figure 3. The correlation coefficients are not very high, 0.230 and 0.270, but highly significant due to the high number of degrees of freedom. However, the significance levels reported in "ecological" correlation studies like the present one must be interpreted with great care, partly due to a phenomenon termed "spatial autocorrelation" - the tendency for neighbouring

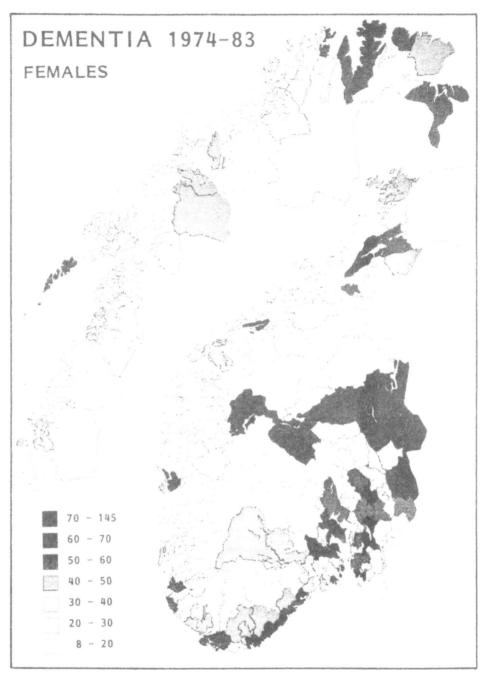

**Figure 2.** *Mean annual, age-adjusted death rates per 100,000 with dementia (ICD 8 290.0 and 290.1 combined) in 193 Norwegian municipality aggregates (10,000+ inhabitants), females, 1974-83. The rates are based on the total number of death certificates (9,085) from which dementia was coded as the underlying or a contributory cause of death.*

250

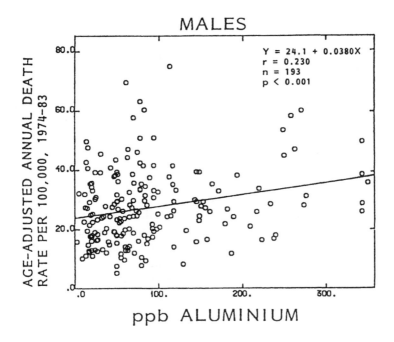

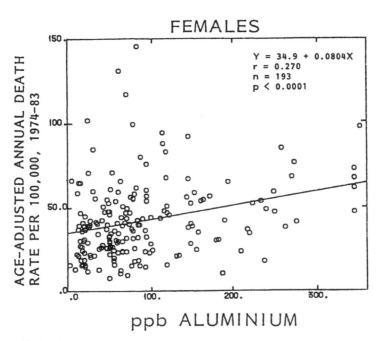

**Figure 3.** *Death rates with dementia (1974-83) versus aluminium in drinking water for 193 Norwegian municipality aggregates (10,000+ inhabitants). Note different scales on the ordinates. The rates are based on the total number of death certificates from which dementia was coded as the underlying or a contributory cause of death.*

geographical units to resemble one another in almost any attribute (Cliff and Ord, 1973). For example, the Al concentration in drinking water in Kristiansand is more likely to resemble that in Mandal (two cities along the southern coast) than the concentration found in Tromso (North Norway). In other words, the assumption of statistical independence is not met, and the number of degrees of freedom on which the significance test is based may be spuriously high. On the other hand, differences in the diagnosis and reporting of dementia (see below) increase the variability in the dementia rates among the different municipality aggregates, thereby generating statistical "noise", which may attenuate the correlation coefficients.

In the present study, disease rates and exposure factors for geographical aggregates are used instead of data recorded at the individual level. Possible errors and uncertainties inherent in this type of study, often termed "ecological" studies, are numerous (Robinson, 1950; Comstock, 1979), and interpretations regarding cause-and effect relationships should therefore be made with great care. For example, if an environmental agent should be involved in the aetiology of AD, it is likely that the *latency period, i.e.* the time lag between exposure and the emergence of the disease, is long. While Al concentrations in drinking water may have increased somewhat in acid rain-sensitive areas during the last decades, the basic *trends* in the geographic pattern are not likely to have changed markedly. *Migration* across municipality boundaries poses more serious problems. Since mortality data are used, movements after onset of the disease and prior to death must also be considered. Since the *weather* in southern and southeastern Norway is better than in the rest of the country, a net influx of new residents of retirement age could be suspected. However, this is contradicted by the fact that the areas with the highest concentrations of Al in drinking water have experienced less increase in population above 70 years within the last 15 years than the rest of South Norway (Vogt, 1986). Severely demented people are often living in institutions, but in Norway, patients in hospitals are residents of the municipality from which they entered the institution, and are therefore coded to this municipality in the cause-of-death statistics. On the other hand, patients in old age homes and psychiatric nursing homes may or may not officially have changed their place of residence. However, these institutions are normally seated in the patient's home municipality, or in a neighbouring municipality, where the Al concentration in the drinking water in most cases is similar to that in neighbouring areas (see above). A special case is constituted by Olaviken psychiatric nursing home in Askoy municipality near Bergen on the west coast. Olaviken is, to the best of the author's knowledge, the only large psychiatric nursing home specialising in the care of dementia patients in Norway. Olaviken receives about two- thirds of their patients from Bergen, and these patients generally change their place of residence. Thus, Askoy, which is a small municipality (population approx. 16,000) has spuriously high dementia rates. Therefore, Askoy and Bergen were pooled to one aggregate before calculating dementia rates (*cf.* Figures 2 and 3). In general, it may be noted that the Norwegian population is rather stationary, compared with those of most other western countries. In 1960, 60% of the total Norwegian population lived in the municipality where they were born (CBS, 1964).

The correlation coefficients for dementia *versus* the percentage of the population living in densely populated areas are 0.288 (males, p<0.00001) and 0.346 (females, p<0.00001), *i.e. higher* than for dementia *versus* Al. However, Al and percentage living in densely populated areas are not strongly *intercorrelated* (r = 0.194, p<0.005). The stepwise regression analysis shows that Al explains an additional amount of 3.1% (males) and 4.3% (females) of the total variability in dementia rates in addition to the variability already explained by the percentage living in densely populated areas. In comparison, Al explains 5.3% (males) and 7.3% (females) of the variability when *not* adjusting for the other variable. The additional explanations of 3.1% and 4.3% are significant by a Fisher test ($F_{1,190}$ = 6.72, p<0.01 (males), $F_{1,190}$ = 9.72, p<0.01 (females). It is again noted that a large part of the variability in the dementia rates is probably caused by differences in diagnosing and reporting of the disease.

The symptom complex of dementia can be caused by more than 60 disorders (Haase, 1977). Of these, it is generally held that AD accounts for 50-60% of the total number of dementia cases, and some additional 10-15% have AD in combination with multi-infarct dementia (Tomlinson *et al.*, 1970; Katzman, 1986). Hence, the diagnosis of dementia is a useful tool in looking for the cause of AD. However, more serious uncertainties are related to the use of *cause-of-death statistics*, *i.e.*, ultimately death certificates filled in by local physicians. In Norway, all death certificates are controlled and coded by the Central Bureau of Statistics. Although most formal errors are eliminated by the control procedures (Glattre and Blix, 1980), dementia will of course not be coded if it is not mentioned on the certificate by the physician filling it in. Different physicians will have different opinions as to whether a disease like dementia can be considered a *cause of death*, especially the *underlying* cause, but also a contributory cause. In the 10 year period studied, dementia was coded from 5,642 (males) and 9,085 (females) death certificates. In 90.4% (males) and 89.1% (females) of the cases, dementia was coded as a contributory cause of death. This means that dementia was coded as the underlying or a contributory cause of death from as much as 2.5% (males) and 4.9% (females) of the *total* number of death certificates in this period. These percentages are not comparable with reported incidence or prevalence rates of dementia. It may be noted, however, that published prevalence rates of severe dementia vary from 1.3 to 6.7% in persons aged 65 and over, with milder impairment in 4.3 to 15.4% (Mortimer *et al.*, 1981; Sulkava *et al.*, 1985). This indicates that a considerable proportion of at least the *severe* dementia cases are noted on death certificates in Norway. However, dementia generally seems to be underdiagnosed in the Norwegian population (Nygaard, 1985). As discussed above, registered death rates with dementia seem to be related to the population density and/or other aspects of the *urban/rural* status. For example, dementia could be less frequently reported in agricultural and/or less central areas, where probably a larger proportion of elderly people live with their families (and possibly a smaller proportion die in hospitals), as compared with the situation in more urban parts of the country. Thus, the major uncertainty of the present study relates to the use of registered death rates with dementia as a measure of incidence rates of Alzheimer's disease. It is possible that the observed geographical association between Al and dementia might be due to differences in diagnosing and reporting of dementia. Epidemiological studies are needed to reveal whether the observed geographical variations in dementia death rates reflect *real* variations in the incidence of Alzheimer's disease. The accomplishment of such studies seems justified by the devastating effects of the disease, its growing burdens on victims, relatives and society (Katzman, 1986), and the increasing (though yet far from conclusive) evidence linking Alzheimer's disease and aluminium in general.

## Acknowledgements

Thanks are due to Bjorn Bolviken and Tor Erik Finne (Geological Survey of Norway (NGU)), Eystein Glattre (Cancer Registry of Norway), Egil Gjessing (Norwegian Institute for Water Research) and Knut Ellingsen (National Institute of Public Health) for fruitful discussions. The water utilities in Norway are thanked for their indispensable help in collecting the water samples. I am grateful to Asbjorn Aase and Erik Nymoen (University of Trondheim) for help and valuable discussions and for permission to use the data base on registered deaths. The chemical analyses were carried out by Magne Odegård (NGU).

This work has been supported financially by NGU, the Royal Norwegian Council for Scientific and Industrial Research, the Foundation of the Technical University of Norway and by the Norwegian Cancer Society.

## References

Candy, J.M., Klinowski, J., Perry, R.H., Perry, E.K., Fairbairn, A., Oakley, A.E., Carpenter, T.A., Atack, J.R., Blessed, G. and Edwardson, J.A. (1986). Aluminosilicates and senile plaque formation in Alzheimer's disease. *Lancet* i, 354-357.

CBS (1964). *Population census 1960, Vol. VIII. Religious denomination - place of birth - citizenship - private car owners - dwelling units with telephone (NOS XII 140)*. Central Bureau of Statistics of Norway, Oslo.

CBS (1974). *Population and housing census 1970, Vol. I. Population by geographical divisions (NOS A 679)*. Central Bureau of Statistics of Norway, Oslo.

Cliff, A.D. and Ord, J.K. (1973). *Spatial autocorrelation*. Pion, London.

Comstock, G.W. (1979). The association of water hardness and cardiovascular diseases: an epidemiological review and critique. In: *Geochemistry of water in relation to cardiovascular disease*, pp.48-68. National Academy of Sciences, Washington DC.

Crapper McLachlan, D.R. and De Boni, U. (1980). Aluminum in human brain disease - an overview. In: L. Liss (ed.), *Aluminum neurotoxicity*, pp.3-16. Pathodox Publishers, Park Forest South, Illinois.

Driscoll, C.T., Baker, J.P., Bisogni, J.J. and Schofield, C.L. (1984). Aluminum speciation and equilibria in dilute acidic surface waters of the Adirondack region of New York State. In: O.P. Bricker (ed.), *Geological aspects of acid deposition*, pp. 55-75. Butterworth, London.

Finnegan, M.M., Rettig, S.J. and Orvig, C. (1986). A neutral water-soluble aluminum complex of neurologicl interest. *J. Am. Chem. Soc.*, 108, 5033-5035.

Flaten, T.P. (1986). *An investigation of the chemical composition of Norwegian drinking water and its possible relationships with the epidemiology of some diseases*, Thesis No. 51, Institutt for uorganisk kjemi, Norges tekniske hogskole, Trondheim.

Ganrot, P.O. (1986). Metabolism and possible health effects of aluminum. *Environ. Health Perspect.*, 65, 363-441.

Glattre, E. and Blix, E. (1980). *Evaluation of the cause-of-death statistics* (Report no. 80/13). Central Bureau of Statistics of Norway, Oslo.

Greger, J.L. (1985). Aluminum content of the American diet. *Food Technol.*, 39, 73-80.

Haase, G.R. (1977). Diseases presenting as dementia. In: C.E. Wells (ed.), *Dementia*, 2nd edn., pp.27-67. F.A. Davis, Philadelphia.

Katzman, R. (1986). Alzheimer's disease. *N. Engl. J. Med.*, 314, 964-973.

Koivistoinen, P. (ed.) (1980). Mineral element composition of Finnish foods: N, K, Ca, Mg, P, S, Fe, Cu, Mn, Zn, Mo, Co, Ni, Cr, F, Se, Si, Rb, Al, B, Br, Hg, As, Cd, Pb and ash. *Acta Agric. Scand.*, Suppl. 22.

Lancet (1985). Acid-rain and human health (editorial). *Lancet*, i, 616-618.

Laxen, D.P.H. and Harrison, R.M. (1981). Cleaning methods for polythene containers prior to the determination of trace metals in freshwater samples. *Anal. Chem.*, 53, 345-350.

Lione, A. (1983). The prophylactic reduction of aluminium intake. *Food Chem. Toxicol.*, 21, 103-109.

Miller, R.G., Kopfler, F.C., Kelty, K.C., Stober, J.A. and Ulmer, N.C. (1984). The occurrence of aluminum in drinking water. *J. Am. Water Works Assoc.*, 76, 84-91.

Mortimer, J.A., Schuman, L.M. and French, L.R. (1981). Epidemiology of dementing illness. In: J.A. Mortimer and L.M. Schuman (eds.),*The Epidemiology of Dementia*, pp.3-23. Oxford University Press, New York.

Nordberg, G.F., Goyer, R.A. and Clarkson, T.W. (1985). Impact of effects of acid precipitation on toxicity of metals. *Environ. Health Perspect.*, 63, 169-180.

Nygaard, H.A. (1985). Senil demens - en underdiagnostisert tilstand? Betydning for situasjonen i somatiske sykehjem. *Tidsskr. Nor. Laegeforen.*, 105, 351-352.

Perl, D.P. and Brody, A.R. (1980). Alzheimer's disease: X-ray spectrometric evidence of aluminum accumulation in neurofibrillary tangle-bearing neurons. *Science*, 208, 297-299.

Perl, D.P. and Pendlebury, W.W. (1986). Aluminum neurotoxicity - potential role in the pathogenesis of neurofibrillary tangle formation. *Can. J. Neurol. Sci.*, 13, 441-445.

Robinson, W.S. (1950). Ecological correlations and the behaviour of individuals. *Am. Sociol. Rev.*, 15, 351-357.

Slanina, P., Falkeborn, Y., French, W. and Cedergren, A. (1984). Aluminium concentrations in the brain and bone of rats fed citric acid, aluminium citrate or aluminium hydroxide. *Food Chem. Toxicol.*, **22**, 391-397.

Slanina, P., French, W., Bernhardson, Å., Cedergren, A. and Mattsson, P. (1985). Influence of dietary factors on aluminium absorption and retention in the brain and bone of rats. *Acta Pharmacol. Toxicol.*, **56**, 331-336.

Slanina, P., French, W., Ekström, L.-G., Lööf, L., Slorach, S. and Cedergren A. (1986). Dietary citric acid enhances absorption of aluminum in antacids. *Clin. Chem.*, **32**, 539-541.

Sulkava, R., Wilkström, J., Aromaa, A., Raitasalo, R., Lehtinen, V., Lahtela, K. and Palo, J. (1985). Prevalence of severe dementia in Finland. *Neurology*, **35**, 1025-1029.

Tomlinson, B.E., Blessed, G. and Roth, M. (1970). Observations on the brains of demented old people. *J. Neurol. Sci.*, **11**, 205-242.

Vogt, T. (1986). *Water Quality and Health - Study of a Possible Relationship Between Aluminium in Drinking Water and Dementia* (Sociale og Okonomiske Studier 61, English abstract), Central Bureau of Statistics of Norway, Oslo.

Weberg, R. and Berstad, A. (1986). Gastrointestinal absorption of aluminium from single doses of aluminium containing antacids in man. *Eur. J. Clin. Invest.*, **16**, 428-432.

Wright, R.F., Dale, T., Gjessing, E.T., Hendrey, G.R., Henriksen, A., Johannessen, M. and Muniz, I.P. (1976). Impact of acid precipitation of freshwater ecosystems in Norway. *Water Air Soil Pollut.*, **6**, 483-499.

# 26 A Possible Relationship between Aluminium in Drinking Water and Alzheimer's Disease in Southern Norway

Tiril Vogt
*The Central Bureau of Statistics of Norway, Research Department, Postbox 8131 Dep., N-0033 Oslo 1, Norway*

## Summary

*A possible link between levels of aluminium in drinking water supplies and the incidence and frequency of Alzheimer's disease, or the onset of early senility, have found some support in Southern Norway.*
*Southern Norway has been divided into five zones according to increasing concentrations of aluminium in lakes, and overall mortality statistics from senile and pre-senile dementia have been compared between the zones. The cause of the rise in aluminium levels in Norway can mainly be attributed to acid rain. The results show a clear relationship between higher mortality statistics and increasing aluminium concentrations.*

## Introduction

*Acid pollution* has in recent years become one of the most serious environmental problems in Norway. How acids are being formed in the atmosphere is now much better understood and there is some scientific concensus emerging over the relationship between pollutive emissions and acid deposition.

*Alzheimer's disease* is the commonest form of dementia in which symptoms of extreme senility appear at a much earlier age than is usual. It is a mysterious disease, which seems at times to have hereditary, infectious and environmental characteristics. Alzheimer's disease is a malfunction of the brain's neurotransmitters - the chemicals that relay signals between nerve cells. The victims gradually lose their intellectual functions, memory and ability to care for themselves, and - in most cases - die within ten years of the first appearance of the symptoms. There is no known cure or effective treatment for Alzheimer's disease. The search for causes of the disease is focused on the tangles of nerve fibres called senile plaques that are found in the brains of the victims.

Acid rain has been tentatively implicated in several brain disorders including Alzheimer's/Alzheimerlike diseases. Drawing on research conducted by a bevy of scientists, the key factor in the diseases may be aluminium (Al) (Maugh, 1984). Usually high levels of Al have been found in the brain lesions of those who suffered from such disorders, although a cause and effect relationship has not been finally proved. Studies in Japan and Guam (Perl, 1985), however, suggest that Al may indeed play a causative role in the disease. The possible link with acid rain is as follows: Aluminium is abundant in the earth, comprising about 5% of the planet's crust. Al is insoluble in water that is either neutral or alkaline. But when the acidity of water - rain, snow, sleet, fog - increases, it begins dissolving Al in lake- bottom sediments, soil, metal pipes used to transport water, and soldering materials used to join sections of pipe.

**Hypothesis**: The hypothesis that has been tested is if there exists a relationship (co-variance) between the concentration of aluminium (Al) in drinking water and the frequency of Alzheimer's/Alzheimerlike diseases in Southern Norway.

## Methods: Use of Indicators

The hypothesis has been tested by use of indicators - both for the intake of Al from drinking water (concentration of Al) and for the frequency of Alzheimer's/ Alzheimerlike diseases.

To establish an indicator of the intake of aluminium through drinking water, Southern Norway has been divided into five geographical zones - by increasing concentration intervals of Al in lakes (see Figure 2, section on *Results*). The pattern of Al in lakes in Southern Norway corresponds rather well with the pattern of Al in drinking water. Other sources of intake of Al - as for example through local foodstuffs or medicine or by using aluminium saucepans for cooking - have not been studied. However, there is no reason to believe that the pattern of intake of Al through other sources in Southern Norway will correspond to the division into the geographical zones that have been used.

Alzheimer's/Alzheimerlike diseases can only be diagnosed by use of specific cerebral studies, which only exceptionally are being carried out for people dying with symptoms of the diseases in Norway. Among all people with senile dementia and pre-senile dementia (age-related dementia) approximately 50-70% are suffering from Alzheimer's/Alzheimerlike diseases, which, however, are not specified in the ICD-8 (International Classification of Diseases, 8th Revision) and thus not given a special code. (They will be included in the ICD-9). Hence mortality per 100,000 population from senile dementia and pre- senile dementia (both sex- and age-specific) coded in ICD-8, has been used as an indicator of the frequency of the occurrence of Alzheimer's/Alzheimerlike diseases in Southern Norway.

*Age-related dementia*: Senile dementia and pre-senile dementia (age-related dementia) are probably the most commonly occurring psychiatric diseases among people of old age, thus being one of the main reasons for psychiatric hospitalisation in Norway. Typical symptoms of the diseases at an early stage are reduced short-time memory, bad time-orientation and simplified and defective speech. At later stages of the disease the patient will also have reduced long-time memory and lost the comprehension of his own identity.

Mainly people above 70 years of age suffer from the disease senile dementia. The term pre-senile dementia is used when the symptoms occur before 65 years of age. "Age-related dementia" comprises both senile dementia and pre-senile dementia in this study.

Age-related dementia will - referring to literature - comprise several diseases being characterised by loss of memory, increasing disorientation as regards time and place and disbehaviour *etc*. The diagnosis of age-related dementia will thus - in addition to Alzheimer's/Alzheimerlike diseases in 50-70% of the cases - comprise conditions of multi-infarct dementia in 10- 20% of the cases. Mixed forms of the two types of dementia comprise the rest of the cases. Multi-infarct dementia is mainly a result of diseases in the circulatory system and hence such diseases and Ischaemic heart disease, Cerebrovascular disease, Hypertensive disease, Diabetes mellitus and Paralysis agitans (Parkinson's disease) have been studied as well.

Of the population in Norway there were approximately 308,000 in the age-group 70-79 years and approximately 142,000 above 80 years by 31 December 1985. Projected population by 1990 is approximately 328,000 (70-79 years) and 162,000 (80 years +), and by the year 2025 408,000 and 191,000 respectively (Norwegian Statistical Yearbook, 1985).

## Data and Sources of Data

Data on drinking water quality have been collected by the Norwegian Geological Survey in 1982-83 (Flaten, 1985). The data refer to 384 larger waterworks in Norway - covering the demand of 71% of the total population in Norway. Of the waterworks, 349 were using surface water and 35 were using ground water as drinking water supply. The survey refers to chemical components of importance to water quality - as for example aluminium. Aluminium in finished Norwegian drinking water is illustrated in the chapter on "Geographical associations between aluminium in drinking water and registered death rates with dementia in Norway" (T.P. Flaten).

Data on acidification of lakes have been collected in the Norwegian project of Acid Precipitation's Effects on Forest and Fish (SNSF-project, 1974). As a part of the project a regional survey of the acidification process in lakes in Southern Norway included data on the content of aluminium in lakes. The areas in Southern Norway most sensitive to acidification are mainly areas poor in calcium and where annual mean pH in precipitation is less than 4.7. This represents a limit for ecological damaging effects in the most exposed areas. In the very south of Norway (corresponding to Zone 5 in Figure 2) the mean pH values in 1985 were as low as 4.2 (see Figure 1).

The Death Register in the Central Bureau of Statistics contains information based on individual death certificates. Deaths resulting from different diseases are classified and coded by using the ICD-8. The register contains - in addition to

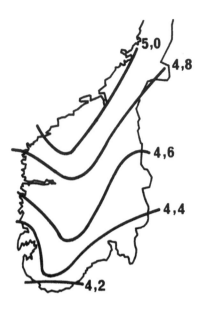

**Figure 1.** *pH mean values in precipitation. 1985. Southern Norway. Source: Norwegian Environmental Statistics, 1987/88.*

cause(s) of death - information on sex, age, time of death, municipality of residence at time of death *etc.* When the medical expert (the doctor) fills in the death certificate, he distinguishes between the *primary cause of death* (the first medical cause of death) and one of four *causes of death* (the complications or additional causes of death). The Central Bureau of Statistics registers and codes both the primary cause of death and - if possible - the three most important of other diagnoses on the death certificates. Annual data based on pre-senile dementia and senile dementia as primary causes of death and as one of four other diagnoses on the death certificate for the period 1969-1983 have been used in this study.

In Norway in 1976 more than 60% of the death certificates contained two or more causes of death, which illustrates the significance of using one of four *causes of death* from the death certificate as a source of data. Norwegian practice shows that doctors regard dementia as a death cause worthwhile reporting in Norway. The diagnoses are being used especially as additional causes of death and, indeed, under-reporting will take place. It is important to point out, however, that there is no reason to believe that Norwegian diagnosis practice will show regional variations. A recent report from NOMESKO (Central Bureau of Statistics, report 80/13) proves only minor regional variations in the diagnoses used on death certificates in Norway. In 90% of the cases the basis for the diagnoses are hospital examinations, the hospitals being the place of death for a rather high percentage of the deaths.

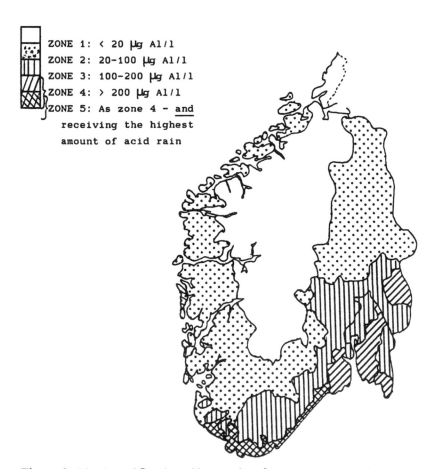

ZONE 1: < 20 μg Al/l
ZONE 2: 20-100 μg Al/l
ZONE 3: 100-200 μg Al/l
ZONE 4: > 200 μg Al/l
ZONE 5: As zone 4 - and
receiving the highest
amount of acid rain

**Figure 2.** *Division of Southern Norway into homogenous zones by Al-concentrations in lakes, Southern Norway.*

The Norwegian death certificates can be regarded as a rather reliable data source, both national and inter-nordic controls have proved very satisfactory results. And accordingly, as regards the death register, as part of the control routines additional information on causes of death are being collected by the Central Bureau of Statistics in approximately one third of all deaths cases.

Data on people with age-related dementia within psychiatric nursing homes in Norway have been collected by the Norwegian Institute for Gerontology in 1982, in a project concerning age- related dementia and institutions in Norway. The survey covered 112 of 118 psychiatric nursing homes in Norway.

*Use of standard population*: Life spans for both sexes have increased during the period of study from 1969-1983. To correct for this where relevant the zones' age structure in this study correspond to Norway's population by 31 December 1975 (standard population).

261

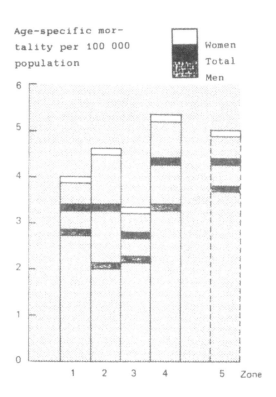

**Figure 3.** *Mortality per 100,000 population, age-related dementia as primary cause of death. 1969-1983. Zones 1-5.*

## Results

The results from the test in this study support the hypothesis of a relation/co-variance between the concentration of aluminium in drinking water and the frequency of Alzheimer's/Alzheimerlike diseases in Southern Norway.

*Highest and increasing mortality in zone 5*: The division of Southern Norway into geographical zones is illustrated in Figure 2. Zone 4 has the highest content of aluminium (Al) in the lakes. Zone 5 constitutes the part of zone 4 that receives the largest amount of acid rain ("the acid rain zone"). The concentration of Al in drinking water within the different zones is illustrated in the figure.

The differences between the zones concerning annual average sex- and age-specific mortality rates (standardised per 100,000 population), have been tested by comparing pairs of mortality rates.

The zone with the highest concentration of aluminium in lakes (zone 5) has the significantly highest mortality (standardised per 100,000 population) from age-related dementia. As well, mortality rates are increasing significantly with

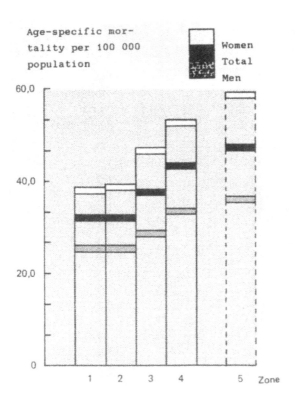

**Figure 4.** *Mortality per 100,000 population, age-related dementia as 1 of 4 causes of death. 1969-1983. Zones 1-5.*

increasing Al- concentrations - from zone 1 to zone 5. The results hold both with age-related dementia as primary cause of death (Table 1, Figure 3) and as one of four causes of death (Table 2, Figure 4) on the death certificate. In the period 1969-1983, a total of 1,575 persons died with age-related dementia as the primary cause of death and 16,341 persons died with the disease as one of four causes of death on the death certificate in the zones.

Development of annual mortality rates over time for age-related dementia as primary cause of death has been tested for people above 70 years of age by use of a non-parametric test and by linear regression. The increase in mortality rate is strongest in zone 5 between 1969 and 1983 - according to both tests, hence the increase in mortality from age-related dementia is strongest in the zone that received the greatest amount of acid rain in Southern Norway within the period of study (1969-1983).

*Mortality from related diseases have different trends compared to the trend for age-related dementia*: Mortality rates from diseases that one *might* suspect have a relation with age-related dementia, show different trends in the period of study

263

**Table 1.** *Number of deaths and mortality per 100,000 population,*
*age-related dementia as* primary cause of death, *1969-1983. Zones 1-5.*

| Zone | | Number of deaths, total | Mortality rate (per 100,000 population) |
|------|------|------|------|
| 1. | T | 235 | 3.4 |
| | M | 102 | 2.9 |
| | W | 133 | 4.0 |
| 2. | T | 580 | 3.4 |
| | M | 182 | 2.1 |
| | W | 398 | 4.6 |
| 3. | T | 261 | 2.8 |
| | M | 111 | 2.3 |
| | W | 150 | 3.3 |
| 4. | T | 499 | 4.4 |
| | M | 192 | 3.5 |
| | W | 307 | 5.3 |
| 5. | T | 166 | 4.4 |
| | M | 69 | 3.8 |
| | W | 97 | 5.0 |

(T = total, M = men, W = women)

(diseases of the circulatory system, ischaemic heart disease, cerebrovascular disease, hypertensive disease, diabetes mellitus and paralysis agitants). In particular, the trends for these diseases decrease from zone 4 to zone 5 during the period 1969-1983, meaning that the trends decrease from the zone with the lowest amount of acid rain to the zone with the greatest amount of acid rain in Southern Norway.

*Mortality from age-related dementia is highest among women*: Higher life expectancy for women than for men might partly explain why mortality rates within the standard population are significantly higher for women than for men above 65 years of age. However, mortality from age-related dementia within all age- groups is higher for women than for men.

*Regional changes in population structure are hardly decisive*: When studying the differences in development of age-related mortality rates between the zones over time, distortion *might* be caused by the increased age of living within different age-groups during the period 1969-1983. However, the data indicate that zone 5 which has experienced the strongest increase in mortality from senile dementia within the standard population, has experienced the least increase in population

**Table 2.** *Number of deaths and mortality per 100,000 population,* *age-related dementia as* 1 of 4 causes of death, *1969-1983. Zones 1-5.*

| Zone | | Number of deaths, total | Mortality rate (per 100,000 population) |
|---|---|---|---|
| 1. | T | 2,216 | 32.4 |
| | M | 937 | 26.1 |
| | W | 1,279 | 38.5 |
| 2. | T | 5,601 | 32.4 |
| | M | 2,233 | 26.0 |
| | W | 3,368 | 39.4 |
| 3. | T | 3,560 | 38.3 |
| | M | 1,456 | 29.4 |
| | W | 2,104 | 47.1 |
| 4. | T | 4,964 | 43.8 |
| | M | 1,918 | 34.3 |
| | W | 3,046 | 53.1 |
| 5. | T | 1,814 | 48.3 |
| | M | 670 | 36.9 |
| | W | 1,144 | 59.4 |

(T = total, M = men, W = women)

above 70 years within the period of study. Accordingly, zone 4 and zone 5 had relatively less people above 70 years of age during 1969-1983 than the other zones. *A relatively high number of people suffering from age-related dementia within "the acid rain zone"*: Figures from a study by the Norwegian Institute for Gerontology show that counties receiving the highest amounts of acid rain have the highest number of people with age-related dementia per 100,000 population within psychiatric nursing homes. The numbers are highest in the counties of Vestfold and Aust-Agder. However, the significance of these figures is difficult to estimate from available data. The number of people with dementia within psychiatric nursing homes *might* be an expression of the state of hospital development (number of beds) rather than the state of the disease itself.

## Further Investigations necessary

This study gives some support to the hypothesis of a positive relation/co-variance between the frequency of Alzheimer's/Alzheimerlike diseases and the concentration of aluminium in drinking water in Southern Norway. However, the data used for the analyses still need to be supplemented before a more firm conlusion can be drawn.

It is important to underline that the co-variance revealed in this study must not be regarded as a proof of an existing linkage between Al in drinking water and Alzheimer's/Alzheimerlike diseases in Norway. However, the study suggests that there exists a link between acid pollution and effects on human health and, accordingly, that further scientific studies should be undertaken. Indeed, it indicates a potential area of serious concern for Norwegian politicians dealing with environment and health!

## Acknowledgements

Mr Eystein Glattre from The Cancer Registry of Norway has acted as medical-epidemiological adviser in this study.

The work has been sponsored by the Norwegian Ministry of the Environment.

## References

Flaten, T.P. (1986). An investigation of the chemical composition of Norwegian drinking water and its possible relationship with the epidemiology of some diseases. Thesis No.51. NTH., Institutt for uorganisk kjemi, Trondheim.

Maugh II, T.H. (1984). Acid rain's effects on people assessed. *Science*, **226**, 21/12.

McCormic, J. (1985). *Acid Earth - The Global Threat of Acid Pollution*. An Earthscan book. International Institute for Environment and Development.

Perl, D.P. (1985). Relationship of aluminium to Alzheimer's disease. *Environ. Health Perspect.*, **63**.

SNSF-project (1975). *Impact of acid precipitation on forest and freshwater ecosystems in Norway*. Fagrapport 6/76 (1980). *Acid precipitation - effects on forest and fish*. Fagrapport 19/80.

The Central Bureau of Statistics of Norway (1985). *Statistical Yearbook*, NOS B530.

The Norwegian Institute for Gerontology (1981). *Age-related dementia and institutional environment*. Per Kristian Haugen. NGI- rapport nr.5.

Vogt, T. (1986). *Water Quality and Health, Study of a Possible Relation Between Aluminium in Drinking Water and Dementia*. Central Bureau of Statistics of Norway, SOS 61.

WHO (1983). *International Classification of Diseases*. 1975- Revision (ICD-8).

# 27 Facts and Fallacies in Dementia Epidemiology

Christopher N. Martyn and E. Clare Pippard
*MRC Environmental Epidemiology Unit, Southampton General Hospital,
Tremona Road, Southampton SO9 4XY, England*

## Summary

*Epidemiological studies to investigate the postulated connection between exposure to aluminium and the development of Alzheimer's disease are essential. We therefore examined the usefulness of existing data on prevalence and mortality as a resource for studying variations in the rate of the disease with time and geography. Unfortunately, methodological differences between prevalence surveys and errors and biases in mortality data are large. No reliable conclusions can be drawn from these data about geographical differences in rates of dementia in England and Wales nor about time trends in the disease.*

## Introduction

Since it is impossible to administer aluminium containing compounds to humans under experimental conditions the hypothesis that links exposure to aluminium with the development of Alzheimer's disease can only be tested by epidemiological techniques. Collection of the sort of data necessary for most epidemiological studies is time consuming and expensive and as a first approach it is sensible to try to make use of information that is already available. Two sources of data relevant to Alzheimer's disease exist; the first is routinely collected mortality data, derived from the information recorded on death certificates by the doctor certifying death; the second is the numerous surveys that have been carried out to determine the local prevalence of dementia. Either might prove to be a resource that would allow investigation of geographical variation in the rate of Alzheimer's disease. Two recent studies from Norway, described in this volume, have used mortality statistics to demonstrate considerable variation in death rates from dementia in different parts of that country. Mortality from dementia was found to correlate positively with the concentration of aluminium in the water supply.

Estimates of mortality derived from death certificates are subject to errors and biases from a variety of sources and preliminary investigations are needed to assess their likely magnitude. This paper describes our investigations of the usefulness of existing data as a way of investigating geographical variation in the prevalence of and mortality from Alzheimer's disease in England and Wales.

## Prevalence Surveys

Recent prevalence surveys of dementia have been well reviewed by Henderson (1986) and by Ineichen (1987). Prevalence rates varied from 2.5% in people aged over 65 years in London (Gurland et al., 1983) to 25% in people over 60 years in the USSR (Sternberg and Gawrilova, 1978). At least part of this variation is due to differences in methodology. There are three main ways in which differences in survey methods are likely to influence the estimate of prevalence: the method of case-finding used, the definition of dementia that was employed and the population that was sampled.

## Case-finding

Surveys which have employed an interviewer to administer a psychometric test or questionnaire to all or a sample of the elderly population in a community are more likely to achieve completeness of ascertainment of cases of dementia than those that have relied on review of cases already known to hospital- based services or general practices. An example where this sort of methodological difference may have had an effect on the estimate of prevalence is easy to find. In Kay's survey in Newcastle in 1970 all people over the age of 65 in the community were sampled; the prevalence of severe dementia was 6.2%. In contrast, in the survey of Adolfsson et al. (1981) that relied on counting cases under institutionalised care the estimate of prevalence was only 2%.

### Definition of dementia
In the absence of generally agreed and workable definitions for dementia and criteria for the diagnosis of Alzheimer's disease, workers are obliged to construct an empirical definition for themselves. Often, these definitions do not attempt to distinguish between different diseases underlying dementia so prevalence estimates of dementia cannot be equated with the prevalence of Alzheimer's disease. It is also obvious that those surveys which aimed to detect cases of mild dementia as well as more severely demented people will obtain a higher prevalence rate than surveys that considered only those cases where dementia was severe. Kay et al. (1985) have discussed in detail the effect that different working definitions of dementia and different psychometric instruments for the detection of dementia are likely to have on prevalence estimates.

### Population sampled
Most surveys have excluded young or middle aged people from the study population because the prevalence of dementia is so low in these age groups. Above the age of

65 years however, the prevalence of dementia increases very fast. In many countries the proportion of very elderly people in the population is growing rapidly. Unless age standardisation has been carried out it is difficult to compare surveys from different countries because the populations surveyed may have had a very different age structure. Another difficulty is that some surveys included only people living at home, while others took a random sample of the whole population over a certain age. Cases of dementia cared for in institutions will be excluded from the former type of survey.

One further problem in the interpretation of prevalence surveys arises because the numbers of cases that were detected in some of them were fairly small. Small numbers affect the precision of the estimate of prevalence. For example, although the prevalence of dementia measured in Hobart was twice that obtained from a similar study in London the confidence intervals around these estimates are such that the possibility that this was a chance finding cannot be excluded (Kay *et al.*, 1985).

Even if we consider only those surveys which employed similar methods of case finding and similar definitions of dementia the inferences that can be drawn from differences in the prevalence of dementia are very limited. No unequivocal evidence yet exists to show geographical variation in rates of Alzheimer's disease.

**Mortality Data**

In England and Wales the Office of Population Surveys and Censuses extracts, from the information recorded on death certificates, the underlying cause of death and codes this diagnosis according to the International Classification of Diseases. This data has recently been used to examine the geographical distribution of mortality from a number of diseases including dementia (Gardner *et al.*, 1984). Mortality from dementia showed considerable variation over different parts of the country. Data derived from death certificates, however, is known to contain a number of inaccuracies. We carried out a series of studies to evaluate whether the apparent geographical variation in mortality from dementia reflected real differences in the rates of the disease (Martyn and Pippard, submitted for publication).

*What proportion of cases of dementia have this diagnosis recorded as the underlying cause of death?*
Using the diagnostic register maintained at the psychogeriatric clinic at Newcastle General Hospital, Newcastle upon Tyne, we identified 197 patients diagnosed as being demented during the years 1980 and 1981. In 1986, when this study was carried out 140 of these patients had died. OPCS were able to provide the death certificates of 137 of these 140. In less than 25% was the underlying cause of death coded as dementia or Alzheimer's disease (ICD Numbers 290 or 331). In only 58% was dementia, Alzheimer's disease or a related diagnosis recorded on either part of the death certificates by the doctor certifying death.

269

*Are patients who do have dementia coded as the underlying cause of death unusual?*

Because so small a proportion of patients who are diagnosed during life as being demented has a diagnosis of Alzheimer's disease or a related condition coded as the underlying cause of death it seemed likely that these patients might be atypical. We therefore examined a sample of death certificates, again provided by OPCS, in which the underlying cause of death had been coded as pre-senile or senile dementia. The sample of 200 death certificates represented approximately one in two of all deaths certified as due to dementia in people under the age of 76 for the year 1978. Two facts of importance emerged; first, the majority of patients had died in hospitals with long-stay beds; second, the most common terminal event by far, was bronchopneumonia.

*Geographical analysis of mortality from dementia*

The mortality data for 1968 - 1978 consisting of extracts from all death certificates for England and Wales were examined. Mortality rates for each sex for each local authority area were calculated from 1971 census data which were grouped according to pre-1974 local authority boundaries. Death rates were expressed as standardised mortality ratios. There were 140 areas which had a significantly higher than average (P<0.05) standardised mortality ratio for men or women or both, and which contained more than five deaths in the 11 year period. These areas were distributed over the country in a remarkably even way; most counties contained two or three areas of high mortality. These areas tended to coincide with the locations of large psychiatric hospitals containing long-stay beds which strongly suggests that the geographical distribution of mortality from dementia is distorted by these institutions. An explanation is found in the practice of OPCS whereby the address of a long-stay hospital is taken as the usual address of the patient after he has resided there for more than six months.

*Time trends in dementia mortality*

The annual numbers of deaths from senile dementia and senility (separate causes of death under the International Classification of Diseases) were examined for the period 1974 to 1984. These data were obtained from published sources (OPCS, 1975 to 1985). Numbers dying from dementia in both sexes rose slowly over the first 10 years but this trend was counterbalanced by a slow fall in the numbers dying from senility. Some increase in numbers of deaths from dementia could have been expected from the increase in the elderly population but it is also likely that changes in diagnostic preference contributed. A striking increase of deaths from dementia occurred over a single year in 1984. In that year OPCS changed its procedures for implementing WHO Rule 3 which deals with how the underlying cause of death is selected from the conditions recorded on the death certificate. Until 1984 where the terminal event was bronchopneumonia the underlying cause of death was coded as bronchopneumonia no matter what else appeared on the death certificate. In 1984, the procedure was changed in cases where in addition to bronchopneumonia, another serious condition contributing to death was also recorded. The second condition was then coded as the underlying cause of death. Because bronchopneumonia is a common terminal event in patients with dementia this change in procedure has had the effect of more than doubling the apparent mortality.

## Conclusions

Methodological differences between prevalence surveys and the deficiencies in routinely collected mortality data, mainly the result of under-reporting of dementia and the distortion produced by regarding long-stay psychiatric hospitals as the patient's usual address, limit the use of these sources of data in the investigation of geographical variation in rates of dementia in this country. It is possible that mortality data collected elsewhere contains fewer errors and biases. But we think that the problems discovered in the data for England and Wales make it obligatory for investigators using similar data from other countries to demonstrate their validity before they can be considered as evidence for a link between aluminium and Alzheimer's disease.

Changes in diagnostic terminology and in the coding procedures used by OPCS render the interpretation of time trends in mortality from dementia almost impossible. Epidemiologists investigating aluminium as an aetiological factor in Alzheimer's disease will have to put aside routinely collected data on mortality from dementia and develop alternative methods of measuring rates of the disease in populations with differing exposure to aluminium.

## References

Adolfsson, R. (1981). Prevalence of dementia disorders in institutionalised Swedish old people. *Acta Psychiatrica Scandinavica*, 63, 225-244.

Gardner, M.J., Winger, P.D. and Barker, D.J.P. (1984). *Atlas of mortality from selected diseases in England and Wales 1968-1978*. Wiley, Chichester.

Gurland, B., Copeland, J., Kurioansky, J., Kelleher, M., Sharpe, L. and Dean, L. (1983). *The Mind and Mood of Aging*. Croom Helm, London.

Henderson, A.S. (1986). The epidemiology of Alzheimer's disease. *Br. Med. Bull.*, 42, 3-10.

Ineichen, B. (1987). Measuring the rising tide. How many dementia cases will there be by 2001? *Br. J. Psychiatry*. 150, 193-200.

Kay, D.W.K., Foster, E.M., McKechnie, A.A. and Roth, M. (1970). Mental illness and hospital usage in the elderly: A random sample followed up. *Comprehensive Psychiatry*, 11, 26-35.

Kay, D.W.K., Henderson, A.S., Scott, R., Wilson, J., Rickwood, D. and Grayson, D.A. (1985). Dementia and depression among elderly living in the Hobart community: the effect of diagnostic criteria on the prevalence rates. *Psychological Medicine*, 15, 771-778.

Martyn, C.N. and Pippard, E.C. Usefulness of mortality data in determining the geography and time trends of dementia. (Submitted for publication.)

Office of Population Censuses and Surveys (1975-1985). *Mortality Statistics (cause) 1974 to 1984, England and Wales*, Series DH2, No.11. HMSO, London.

Sternberg, E. and Gawrilova, S. (1982). Uber klinisch- epidemiologische untersuchungen in der Sowjectischen Altenpsychiatrie. *Nervenzart*,49, 347-353 (cited in Ineichen, 1987).

World Health Organisation (1967 and 1977). *International Classification of Diseases 1965 and 1975*. 8th and 9th revisions. HMSO, London.

MONOGRAPH SERIES

ENVIRONMENTAL GEOCHEMISTRY AND HEALTH

# Geochemistry and Health

## Proceedings of the Second International Symposium

*Edited by*
Iain Thornton

## 8  A Study of Environmental Geochemistry and Health in North East Scotland

P.J. Aggett, C.F. Mills, A. Morrison, M. Callan, J. Plant, P.R. Simpson, A. Stevenson, I. Dingwall-Fordyce, C.F. Halliday

PLATE CAPTIONS

**Relationship between Copper Deficiency in Cattle in North East Scotland and digital imagery of the geology and geochemistry of minus 150$\mu$m stream sediments from the Geochemical Atlas of Moray-Buchan prepared by the British Geological Survey**

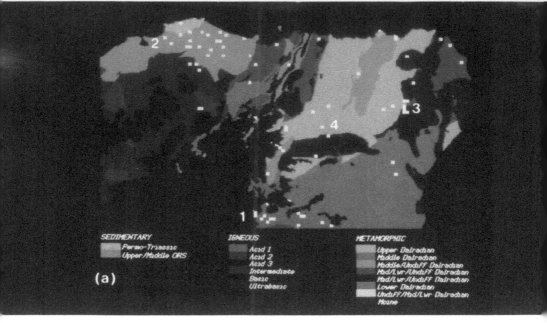

**(a)**

(a) Copper deficiency incidents in relation to digitised map of the solid geology. Disease incidents are apparently diffusely scattered over Dalradian rocks but correlate well with the following geological features: Area 1. Acid igneous rocks of the Aberdeenshire Granite Batholith; Area 2. Upper and Middle Old Red Sandstone in the Laoigh of Moray; Area 3. Upper Dalradian metamorphic rocks near contact with Haddo basic intrusive; Area 4. Upper Dalradian metamorphic rocks near contact with Insch basic intrusive.

Copper concentrations : ppm
Low                              High

**(b)**

(b) Copper deficiency incidents in relation to copper geochemistry. No apparent relationship exists between copper geochemistry and the reported incidents of copper deficiency disease. This accords with previous results obtained from data on crop and soil composition.

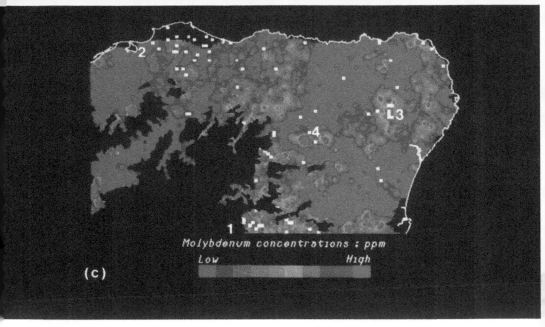

(c) Copper deficiency incidents in relation to molybdenum geochemistry. Area 1. Has high incidence of copper deficiency with anomalous molybdenum geochemistry. Area 2. Has a high incidence of copper deficiency but shows no clear association with high molybdenum geochemistry. Area 3. Has acute copper deficiency disease problems which coincide with a high molybdenum anomaly. Six locations have been identified from veterinary investigations in this area which all lie within the high molybdenum anomaly and also have very high herbage molybdenum anomaly.

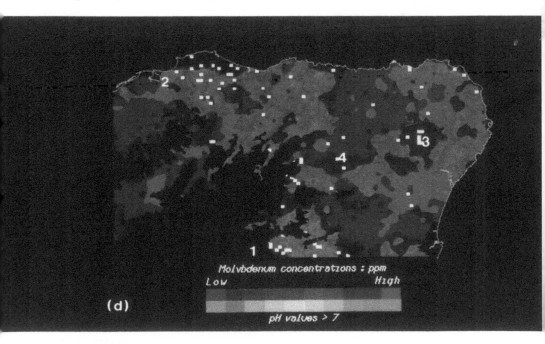

(d) Copper deficiency incidents in relation to molybdenum geochemistry and stream pH greater than 7 (lighter tones of concentration scale). Only 5 out of a total of 66 copper deficiency incidents occur where molybdenum geochemistry is below 2 ppm or stream water pH is below 7.

9 781138 558892